FIBER GLASS

FIBER GLASS

J. Gilbert Mohr
Johns-Manville Sales Corporation
Process Development Division, Waterville, Ohio

and

William P. Rowe
Johns-Manville Sales Corporation
Filtration and Industrial Minerals Division, Waterville, Ohio

VAN NOSTRAND REINHOLD COMPANY
NEW YORK   CINCINNATI   ATLANTA   DALLAS   SAN FRANCISCO
LONDON   TORONTO   MELBOURNE

Van Nostrand Reinhold Company Regional Offices:
New York Cincinnati Atlanta Dallas San Francisco

Van Nostrand Reinhold Company International Offices:
London Toronto Melbourne

Library of Congress Catalog Card Number: 77-18264
ISBN: 0-442-25447-4

Manufactured in the United States of America

Published by Van Nostrand Reinhold Company
135 West 50th Street, New York. N.Y. 10020

Published simultaneously in Canada by Van Nostrand Reinhold Ltd.

15 14 13 12 11 10 9 8 7 6 5 4 3 2

Library of Congress Cataloging in Publication Data

Mohr, John Gilbert, 1913–
    Fiber glass.

    Bibliography: p.
    Includes index.
    1. Glass fibers. I. Rowe, William P., joint
author. II. Title.
TP860.5.M63   666'.157   77-18264
ISBN 0-442-25447-4

Foreword

Previous editions of the Handbooks of Fiber Glass and Reinforced Plastics in which one of the authors was involved were so well received and were so useful as a standard work on its ever-expanding technology that it was determined to prepare yet another volume which would set forth the current status of technology and product applications for the entire field of fiber glass. Because of the current need for energy savings, the new volume stresses the wool and insulation side of the fiber glass industry. Only 26% of the houses in the cooler climates of the United States are now insulated and there is, of course, a continuing demand in the warm areas for insulation for air conditioning and solar collectors. These facts make those portions of this book dealing with insulation of particular interest to technical persons, engineers, and management.

A book of this type is particularly useful in providing a wealth of basic information on which new fiber glass products as yet unknown can be developed.

The authors deserve the continuing commendation of all who are interested in fiber glass technology.

JOHN A. McKINNEY
President
Johns-Manville Corporation

Preface

While glass and ceramics materials are almost as old as civilization itself, the science and technology associated with these two materials had fairly recent beginnings—sort of like the proverbial dot on the endless piece of string. In fact, quite a lot has been found out about the material glass just since 1900. The atomic structure of glass was studied intimately, reasons for strengthening by heat tempering were delineated, and beneficial manipulations of crystalline phases were accomplished. At the same time, ingenious methods were devised for large-scale melting operations and associated high-rate producibility of bottles and holloware, flat glass and scientific glass, and many other products generally benefiting mankind's lot. The whole record is described in a fine book, "Revolution in Glass Making," by Scoville.

An erudite group of the scientists and technologists mainly associated with and responsible for the great success of these ventures grew up and banded together, becoming justifiably proud of their accomplishments and generally accepting the status quo. The word "glass" was uttered with hushed reverence. At one time many of them assembled on a wild and windy shore of the Chesapeake Bay and attempted to duplicate the Phoenecian discovery of glass described in Chapter 1.

As a young upstart graduate student circumnavigating the fringes of this select circle, your author continued to shock some of the senior members with such questions as, "With all we know about its structure, why can't we produce a glass that will bend?" and "How come glass won't behave like plastic materials?" The answer was always the closed-door kind that went something like, "It's not the nature of the material."

The chronology of these sparring matches coincided almost precisely with the commercialization of fiber glass and the start of the industry. Fiber glass was welcomed as a material with a completely different set of properties that made possible new products with completely new performance capabilities. Glass in the fibrous form *would* bend. When combined with plastic resins, a completely new and different material resulted which had highly desirable properties previously unattainable.

The first applications of fiber glass products were in filtration, but before any markets were firmly established, World War II was precipitated, and like many others, fiber glass was called upon as a strategic material. Nonmagnetic land mines, jettisonable fuel tanks, plus highly efficient thermal and sound insulation for military aircraft and shelters helped win the war.

At the close of hostilities, it became necessary to familiarize other scientists, engineers, and designers with the properties of fiber glass so that the unique products developed could be converted to peacetime usage, and so that the production capability built up during the war period would not be completely wasted and many jobs sacrificed.

So the parent company, Owens-Corning Fiberglas Corporation, set out to acquaint the American public with all the great things fiber glass could and would do. In addition to granting licenses to produce fiber glass, they organized and staged an elaborate road show which extolled the benefits of fiber glass and revealed its many unique and unusual properties. This road show toured the country for several years and truly made industrial history.

Every conceivable yet practical industrial, economic, and domestic use for fiber glass was thought of and its advantages clearly demonstrated. Its acoustical qualities were demonstrated by placing a large, loudly ringing alarm clock inside a 6 or 8 in. diameter fiber glass pipe, where the noise became barely audible. The thermal performance was aptly shown by wrapping a frozen one-quart ice cream package in a fiber glass insulation blanket, and subjecting it to the heat of a baking oven in the same compartment with a pie being baked. At the conclusion of the baking time required for the pie, both were removed from the oven, and pie a la mode served, the ice cream still being in the frozen state when unwrapped.

The great strength of continuous-filament fiber glass was demonstrated when an all-American, heavyweight football player touring as a member of the show group chinned himself on a bar suspended by thin bands of fiber glass reinforced tape. Its strength was also made very evident when the strongest muscleman in the audience was invited to whack away with all his might using a heavy hammer alternately at a square of heavy metal sheet (which crumpled from the blow) and an equivalent size panel of fiber glass reinforced plastic laminate (which completely resisted the blow and sent the hammer recoiling upward with equal force).

Also presented in this demonstration were many other interesting and utilizable properties of fiber glass which proved its efficacy and practicability in many applications. The road show did its job, because fiber glass has become accepted by engineers, scientists, industrial entrepreneurs, and by us, the purchasing public.

This volume is a testimony to that great growth, but more so it is a theater or stage upon which the further accomplishments in fiber glass since those early days will be enacted. Like any book, however, the words are frozen in time and cannot be changed. What will change and continue to progress even more favorably will be better applications and performance, and new areas and fields of usage for fiber glass. Combinations of materials and circumstances made fiber glass possible, but people made progress happen. Becoming involved technically and economically with fiber glass has spelled success for many persons. In case you are still in the audience, please believe that there is still plenty of room on the stage!

J. GILBERT MOHR
WILLIAM P. ROWE

Contents

FIBER
GLASS

Chapter 1 | *Introduction*

In the annals of industrial progress, man has continually applied pressure to force back the boundaries of nature and wrest her secrets. Mother Nature usually resists in the most stubborn fashion, but once in a while gives up a whole parcel of her mysteries at one time, yielding to the pressure as if to grant begrudgingly, certain options, then retrenching to the next level of resumed resistance.

Examples are manifold in construction, mining, metals, and many other fields, including the field of glass and glassmaking.

Reputedly, Mother Nature yielded the original secret of glassmaking to a Phoenician sailor bent on building a fire on a sand beach using blocks of nitre (potassium nitrate) from the hold of his ship as a fireplace. The sand piled around the blocks stacked to hold cooking vessels reacted with the nitre, as a eutectic was formed due to the heat of the fire. When the shining trickle of viscous glassy liquid caught his eye, he tried again and again until he could produce the material on a reliable, duplicative basis.

Following, jewels, beads, and amulets were produced and became one of the important trading media in the then civilized world. Next the blowpipe, then larger melts, and a greater quantity of decorative and utilitarian items of glass—all learned after tiring effort and successive failures.

In our own century, automated machine production of glass tumblers, bottles, and pressedware got its start and flourished as a result of a great series of inventions between 1880 and 1920.

The scientific reasons for heat-tempered (strengthened) glass, long a laboratory curiosity, were determined in this century. With process control came greatly expanded utility of stronger glass for architectural, automotive, and other uses.

1

Fiber glass, also an early glass oddity, expensively produced by crude winding or pulling machines for centuries, came into its own, and joined the great industrial expansions known to the world.

First appearing as a fine decoration on Syrian and Egyptian glass vessels and amphora, then as a threadlike incorporation in the beautiful creations of Venetian glassblowers, fiber glass was subsequently made by both French and German commercial glass producers in the early to middle 1700's.

In 1892, along with other items determined to put glass, including the fibrous form, on an everyday common usage basis, a fiber glass dress and parasol were exhibited at The Columbian Exposition in Chicago. The two exhibitors were responsible for the greatest single revolution in glass, and this event competes with or stands high on the list of all-time industry revolutions anywhere in the world.

The present fiber glass industry resulted when two large American glass companies instituted a great joint research effort to actually find methods of producing the fibrous product in an economical manner. In the melee our *Madre Natura* gave up many secrets in one fell swoop. Continuous filaments, the then known form, were produced in strands from a heated platinum box or crucible containing a multiplicity of orifices instead of the single hole, which their predecessors had used.

Blown glass fiber was discovered also during this series of researches. It was noted that a bead blown off the end of a glass rod thrust in a jet flame carried a fine thread of glass with it due to the high surface tension and viscosity of the hot glass. This mass of bulky glass, it was determined, had better thermal insulating qualities than the slag wool manufactured at the time. In the present era, sales of fiber glass for insulation and associated applications have, in the time of only one generation, increased to 2.1 billion pounds per year, and those for the continuous-filament to 1.1 billion pounds per year (1977). Growth rates of 10 to 15% per year are still being realized.

This book is intended to systematically tell of the almost endless stream of utilitarian and life-easing products that have resulted from the discovery of mass production of these two types of fiber glass: bulk or blown fiber for insulation and allied applications, and continuous-filament, or reinforcing fibers.

Since the end applications are distinctly different, the two general fields, blown and continuous fibers, are treated separately herein, Methods of manufacture plus resultant products and end applications for each are detailed and copiously illustrated. Also given are technological particulars responsible for the success of each application.

Chapter 2 | # Methods of Forming Fibrous Glass for Insulation and Mat Products

INTRODUCTION

Although many natural materials were used in the past by man, answering his instinctive urges to prevent heat loss from or entry into his dwellings, no material in modern technology has satisfied the all-around requirements as has fiber glass.

The precursor of all glass insulation, acoustical and other fibrous building products was mineral or slag wool. The first mineral wool on record was produced in Wales about 1840, and it was used to prevent heat loss from boilers and steam pipes. This product was difficult to handle because of the coarse fibers and "shot" included (minute glassy beads with sharp points or tips). However, its usage flourished especially in America, with a large segment of the industry centering around Alexandria and Richmond, Indiana. Large producers were Johns-Manville Corporation and National Gypsum Company.

Large-scale production of glass fiber on a commercial basis dates from 1932. The first product was a coarse-fibered air-cleaning filter made from randomly oriented fibers blown from a jet and laid down in a jackstraw pattern. Finer fibers for insulation followed.

The Owens-Corning Fiberglas Corporation was formed in 1937. Mass-produced continuous-filament fibers also came on stream at that time (see Chapters 4 and 5). Glass fiber products for building, acoustical, filtration, and other applications followed in rapid succession.

MELTING GLASS FOR FIBERIZATION

Products of fibrous glass for all of the three main areas of application, low, ambient, and high temperature insulations, and all other fields must start with a glassy melt. It is of preliminary interest to discuss briefly the three major melting systems in use: the cupola, the glass tank furnace, and electric melting.

The Cupola

A cupola is defined as "an upright cylindrical furnace used for melting cast iron." The cupola adapted for production of inorganic fiber (mineral or slag wool, or high temperature fiber) answers this description (see Fig. 2-1). Blast furnace slag, or layers of slag, limestone, and coke were layered before firing and then fed in continuously after the melt started. Once induced, the fire continuously maintained its heat due to the burning coke layers. Temperatures of

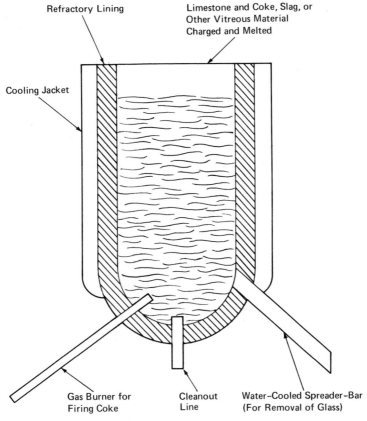

Fig. 2-1. Cupola.

2000 to 2200°F were reached.   The glassy melt exuded from a suitable refractory-lined opening and was conducted to the fiberizing mechanisms.   Capacities of cupolas vary, with the largest melting 10 tons per day.

The Glass Tank Furnace

Glass tank furnaces move glass horizontally through three stages or zones: (1) melting (of raw batch charged), (2) refining, and (3) working.   Refining and working zones successively perform the functions of homogenization (uniformity of melt and release of bubbles) and lowering of the temperature to the proper point so that the glass is at the correct viscosity for processing.   A typical glass tank furnace is shown schematically in Fig. 2-2.

Glass produced through a tank furnace may be either fiberized directly or made into marbles for subsequent feeding to remelting units with specialized fiberizing capability.   Capacities for fiber glass production range up to 70 or 80 tons per day.

Electric Melting

Adaption of electric melting to ceramic products was made approximately from 1935 to 1940 when ternary-electrode furnaces were used to smelt aluminum oxide in a large crucible.   The raw unmelted Al_2O_3 material around the periphery of the crucible acted as its own insulator and temperatures well in excess of 3700°F (2100°C) were

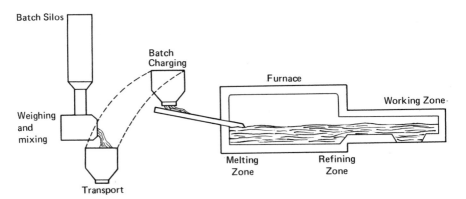

Fig. 2-2.   Schematic Diagram of glass tank furnace.

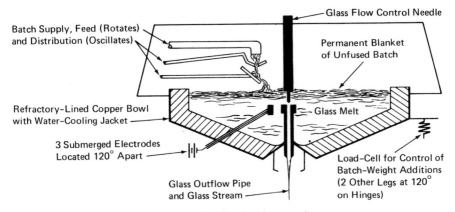

Fig. 2-3.   The Pochet furnace.*

reached.   The melted material was poured out into molds to form "fused cast" refractories for glass tank blocks, etc.   A French adaption of use of a ternary-electrode furnace for melting glass, the Pochet furnace, is shown in Fig. 2-3.   Glass from this furnace could possibly be delivered into channels and forehearths which could then reduce it to the proper temperatures for conversion into fibers.   The rate of melting per unit area of contact surface is much greater than that for any other type of glass melting.

PROCESSES FOR FIBERIZATION

The several processes for converting melted glass into the fibrous state are described as nearly as possible in chronological order of their inception and development.   As previously indicated, fiberizing mechanisms have been developed to use glass from either the primary melting unit, or to use marbles which must, of course, be reheated. The direct-melt processes have an economic advantage in that additional fuel is not required for remelting the marbles.   The marble process is more adaptable, however.   Marbles can be shipped to many differently located fiber-producing plants from a central point.   The marble remelting units are small and separate, and are more easily maintained and replaced.   Also, with an operation geared to use of marbles, the glass composition of the marbles may readily be

*U.S. Patent Nos. 3, 147, 328; 3, 376, 373; and 3,429,972.

changed for the purpose of producing different types of fiber glass products.

There is nothing magical or mysterious about selection of marbles for processing into fibers. Convenience is the criterion. Being round, they move downhill through graders, etc., by the action of gravity. They also may be manufactured at a high rate of speed, and degree of annealing is not highly critical.

Disadvantages of marbles are that they do not always show glass defects (chemical imperfections such as cords and stones), neither does a manufacturer receive 100% recovery since many marbles are defective (outsized, etc., or broken in shipment).

As discussed, coke, gas, oil, or electricity may fuel the basic glass-melting units. Marble processing devices, however, are gas fired, or consist of clay pots heated by induction, or of platinum bushings brought up to glass-melting temperatures by electrical resistance heating.

Mineral Wool

Figure 2-4 illustrates schematically the fiber producing facilities of the Ohio Insulation Company that operated in Toledo between 1935 and 1945.* Layers of coke, slag, and limestone were charged to the cupola, fire was started with a flame on a long lance or probe, and

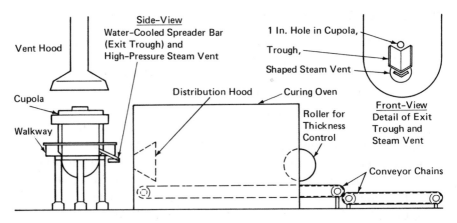

Fig. 2-4. Mineral wool process.

*Personal Communication, Mr. Carl D. Baker, Toledo, Ohio.

the melted fraction channeled down a V-shaped trough. Fiberization was by steam through jets at the bottom of the trough. A liquid binder was applied at the bottom of the trough, atomized and remained on the fiber, and was cured by heat of the steam as the material entered the oven. The binder is necessary to preserve mat integrity and to provide resilience. Once the melt was under way, additional layers of coke and limestone were charged continuously to the top of the cupola by inclined conveyor. A weekly shutdown and cleanout was necessary for both cupola and curing oven. The material was produced either in rolls, batts, or bulk (by picker action). A large percentage of coarse fiber and shot were coincident with the fiber attenuation.

Steam-Blown Process

Another process adaption for rock-wool was the technique of blasting steam jets into downward-flowing free-streams of melted vitreous material (see Fig. 2-5). Originally used for probably the major portion of rock wool, this method was first employed to attenuate glass into the fibrous state. It also comprises, or gave impetus to methods

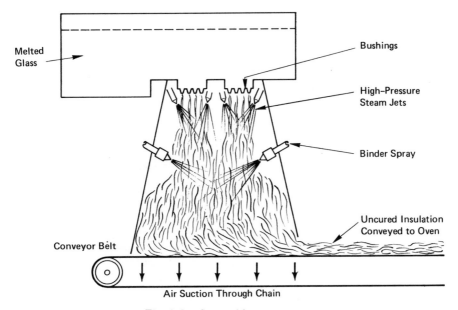

Fig. 2-5. Steam blown process.

now used to form high temperature fibers from alumina-silica and like refractory melts.

Again, the fiber mass so produced contained unfiberized "shot," and consequently was limited in its ability to resist flow of heat.

Flame Attenuation Processes

Adaptable either to direct glass supply from a furnace source, or to remelting of marbles, this process was developed by Owens-Corning Fiberglas Corporation in the middle or late 1940's. It consists of first drawing down finite-sized "primary" fibers approximately 1 mm in diameter, then aligning them in an exact, uniformly juxtaposed array into the jet flame blast issuing from an internal combustion burner. Here the fiber thinning and lengthening takes place immediately, with fine, long fibers produced in the mass. (See Fig. 2-6A and Fig. 2-6B, Toration Process.) It is one of those "you-wouldn't-believe-it" experiences to watch, using a protective infrared-absorbing

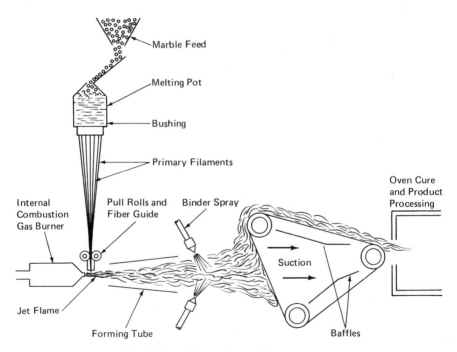

Fig. 2-6A. Flame attenuation process.

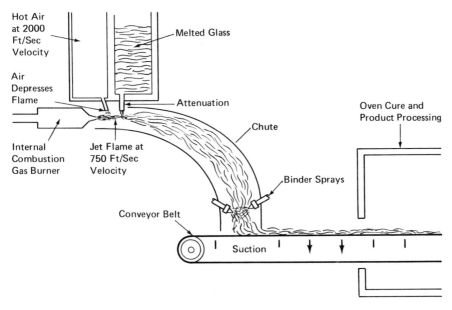

Fig. 2-6B. Toration process.*

glass, the ends of the primaries dancing up and down in the flame, with long tails of the finer attenuated fiber flowing hurriedly away, then to reach into the stream with an iron tool at point of formation and collect a portion of the fine, cottony mass for visual inspection.

Binder is sprayed on to establish mat integrity and resiliency, and cure is accomplished through an oven while thickness and density are controlled by adjusting the level of perforated upper "flights."

This process provided the capability of actually varying and closely controlling fiber diameter over a wide range. In fact, fibers down to 0.000002 in. (0.05 μ) diameter can be produced commercially. Average diameters are 0.00005 to .0001 in (AA to B-fiber.)

Spinning Process

This process was developed in 1955 by Johns-Manville Corporation to provide improvements over the steam-blown process. Similarly, the glass or mineral material is cupola-melted. However, the glassy

*French patent Nos. 2, 223, 318, and U.S. Patent Nos. 3, 885, 940 and 3,874,886.

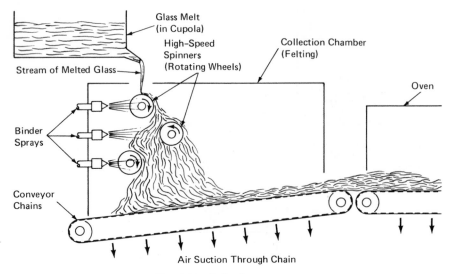

Fig. 2-7. Spinning process.

stream free-falls vertically onto a series of rapidly rotating wheels which induce attenuation by casting the hot glass into the chamber space and onto the other wheels for further fiberizing (see Fig. 2-7).

Fibers so produced are finer and generally longer in length than steam-blown fibers, hence provide end products with superior insulating values.

Rotary Process

This is the latest in the line of improvements in bulk (wool) fiber production, and provides material with fiber diameter and length equivalent to those for the flame attenuation process, but with considerably greater output per fiber producing unit.

The glass stream from the melting unit flows by a free-fall drop into a hollow cylindrical unit with holes in its vertical sidewall edge (see Fig. 2-8). The rapid mechanical rotation of the cylinder forces the glass stream to break up and be exuded horizontally from the holes. Once outside, the glass streams are then further attenuated by action of high-pressure, downwardly-directed air jets located peripherally around the spinning cylinder. As usual, fibers are col-

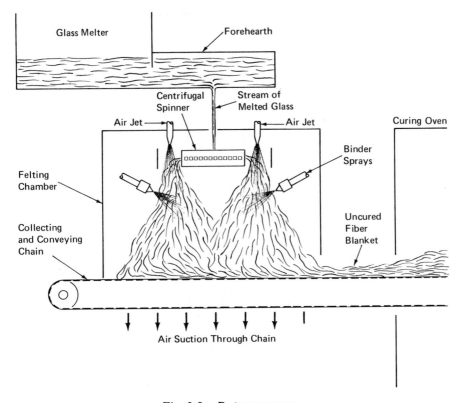

Fig. 2-8. Rotary process.

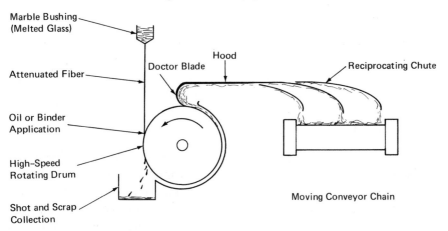

Fig. 2-9A. Schuller process.

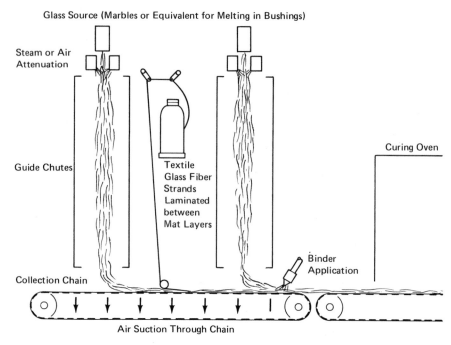

Glass Source (Marbles or Equivalent for Melting in Bushings)

Steam or Air
Attenuation

Guide Chutes

Curing Oven

Textile
Glass Fiber
Strands
Laminated
between
Mat Layers

Binder
Application

Collection Chain

Air Suction Through Chain

Fig. 2-9B. Bonded industrial mat process.

lected after spraying with binder, and cured in an oven. Shot content is low, and the usual diversity of products may be fabricated.

Mat Processes

Two extant mat processes provide fiber glass mat products directly from glassy melts. These mats have wide industrial usage not necessarily applicable to thermal and acoustical problems, but are not in the continuous-filament category, hence will be included here.

Schuller and Associated Processes

Primary hot glass filaments from a platinum resistance bushing are attenuated into small-diameter, noncontinuous fiber by a large, rapidly spinning drum. Shot and other offal are initially rejected out of the product and the good fiber is removed from the drum via stationary doctor blade. It is then conveyed by chute or other means to a moving belt where the desired products are determined by binder, control of thickness, density, etc. (see Figs. 2-9A and 2-9B).

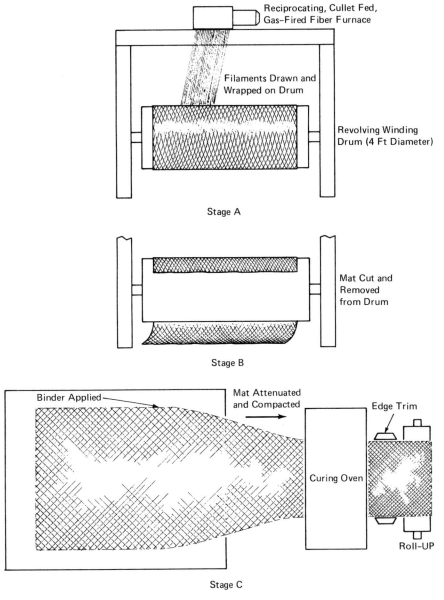

Fig. 2-10.   Mechanical attenuation of glass fibers for mat production.

Mechanical Attenuation (Modigliani Process)

A direct-melt (if large enough) or remelt furnace (cullet or marbles) is arranged to reciprocate over the top of and in parallel with the longitudinal axis of a large revolving drum (see Fig. 2-10). Fibers are wound on the drum in the desired thickness. If a filament beads out of one of the bushing tips, it may be refed into the mat with a quick downward motion without disturbing the winding fiber bundle. Mats are cut off the drums and attenuated sideways to form a large complement of interesting and useful fiber glass mat products.

Fig. 3-3. SEM photo (300X) of fiber glass Micro-Fiber® of 1.5 μ mean diameter.

bubble release from the melt, long working range, and facility of fiberization are important.

In the room-temperature condition for end-use application, the fiber composition should possess good chemical durability and resistance to water attack because of the much larger surface area exposed. It should also accept binder properly, should have a high mechanical strength and lack of friability.

Typical chemical compositions for mineral wool, insulation-type, high temperature fiber glass are shown in Table 3-1. Important parameters for evaluating and controlling glass compositions are liquidus temperature (point of initial crystal formation out of the melt upon cooling), softening point (temperature at which glass, a thermoplastic, softens and flows under its own weight), density (weight per unit volume determined after controlled thermal history or annealing), rate of flow at the fiber-forming temperature (a viscosity test), and seed count (either entrained or dissolved gases being released or incomplete melting). Naturally, chemical analysis by any of several reliable methods is essential for control of both raw glass-batch materials and finished melted glass. Periodically it is also advisable to evaluate the finished glass for its chemical durability. This is

TABLE 3-1.   Formulations for Insulation-Type Glasses.

	Formula for typical mineral or slag wool	Typical fiber glass insulation composition	Typical high-temperature fiber composition
SiO_2	50	63	50
Al_2O_3	10		
Fe_2O_3	1	} 6	} 40
CaO	25	7	6
MgO	14	3	4
Na_2O	—	14	—
K_2O	—	1	—
B_2O_3	—	6	—
F_2	—	0.7	—

done by measuring weight loss after exposure of fibers of a known, closely controlled filament diameter to water and to acids and bases of a predetermined normality.[1]

Fiber Diameter

This is the most important basic factor as regards specific performance for fiber glass and associated materials, since almost all major end-use behavior is determined by fiber diameter. Generally product cost increases proportionately with the necessity to create finer filament diameters.   The finer-fibered products will do most of the things that those with coarser fibers will, and even more. Hence end-use requirements should be carefully assessed—if only cold cuts are required, it is not necessary to pay for prime-grade steak.

Filament diameters and ranges applicable to all fiber glass production are presented in Table 3-2.

In quality control of fiber sizes for a blown fiber glass production operation, diameters are measured by resistance to air flow using a testing device developed by the Sheffield Micronaire Division of Bendix Corporation (see Fig. 3-4).   Originally intended for evaluating cotton, this device may be recalibrated for glass fibers. Small standard cylinders containing a weighed mass of fibers of known diameter and range are used to set or produce one specific air flow rate in the test unit.   Following, a weighed portion of an unknown fiber sample is loosely packed into a likesized test cylinder, inserted, and

TABLE 3-2. Filament Diameter Conversion Chart.

	INCHES		MICRONS	
	Min.	Max.	Min.	Max.
AAAAA	.000002	.000008	.05	.20
AAAA	.000008	.00002	.20	.50
AAA	.00002	.00003	.51	.76
AA	.00003	.00006	.76	1.52
A	.00006	.00010	1.52	2.54
B	.00010	.00015	2.54	3.81
C	.00015	.00020	3.81	5.08
D	.00020	.00025	5.08	6.35
E	.00025	.00030	6.35	7.62
F	.00030	.00035	7.62	8.89
G	.00035	.00040	8.89	10.12
H	.00040	.00045	10.12	11.43
J	.00045	.00050	11.43	12.70
K	.00050	.00055	12.70	13.97
L	.00055	.00060	13.97	15.24
M	.00060	.00065	15.24	16.51
N	.00065	.00070	16.51	17.78
P	.00070	.00075	17.78	19.05
Q	.00075	.00080	19.05	20.32
R	.00080	.00085	20.32	21.59
S	.00085	.00090	21.59	22.86
T	.00090	.00095	22.86	24.13
U	.00095	.00100	24.13	25.40

1 Micron equals .00003937 inches
(39.37 millionths of an inch)

its resistance to air flow measured. The mean fiber diameter of the test sample is smaller or greater than the control standard depending upon whether the sample offers, respectively, more or less resistance to the flow of air.

One difficulty with this measuring system is that the extremes, or degree of fiber diameter distribution under and over the nominal value (3-Σ limits) can not be accurately determined. Nevertheless the method has provided the industry with a good, practical, and duplicatable control of fiber diameter.

Diameters down to 1 μ may also be measured optically at 1,000 diameters using an accurate projection microscope with calibrated

Fig. 3-4. Quality control technician loading weighed fiber glass sample into a standard cylinder for filament diameter measurement by air flow technique.

screen. This system is more laborious, requires excellent equipment and precise operator technique, but provides extremely accurate results.

Binders

Raw glass fiber in any form, blown bulk or continuous, is brash and easily fragmentized. This is because self-abrasion induced by any kind of motion or rubbing action causes surface defects. These in turn reduce flexural, tensile, and other mechanical strength parameters. The adage is also true with fibers as with other forms of glass that glass is only as strong as its surface.

Consequently, a family of various types of "binders" for mineral and glass wool products has been developed. Applied from 5 to 25 wt% depending upon application, binders are based mostly upon phenol-formaldehyde resins for bonding; they also are formulated to include melamine resins, silicone compounds for water repellency, soluble or emulsified oils for lubrication, wetting agents for control of surface tension, and extenders or stabilizers.

The phenol-formaldehyde resins used are of the strong-base resole (one-step) type, and are water-soluble with a specified dilutability or tolerance of up to 25 volumes of water. Fire-retardant additives are usually reacted in the resin formulation. The resins must be refrigerated prior to use but have fairly long-term (24 hr) stability in the mixed-binder state. The phenolics cure (polymerize) on the glass by chemical action induced by heat (350 to 500°F in the wool; up to 700°F ambient in the curing ovens). Resin age, pH, percent solids, and degree of cleanliness are important factors in cure.

In the binder formulations used, the end results justify the care and difficulties required in handling. When sprayed on immediately after fiberization or attenuation, the resin accumulates in droplets around the fiber. Then during cure, these droplets become transpelled along the fibers, reaching fiber junctures or simply flattening out along the fiber. Hence both protection against abrasion and resiliency for the final product are provided. The deposition and flattening-out of resin droplets along fiber surfaces, and also accumulations at junctures of two or more fibers are clearly visible in the SEM photomicrograph (Fig. 3-2).

Raw phenolic resins may be tested for cure temperature and time on a standard cure plate.[2] (See Fig. 3-5.) Degree of cure of resin ap-

Fig. 3-5. Utilization of a standard cure plate for determining the setting temperature and corresponding time rate for binder resins in fiber glass wool products.

Fig. 3-6. The method of burn-off at an intermediate temperature which removes the organic binder but does not change the glassy mass. Percent weight loss is then calculated.

plied to glass wool products may be evaluated by color (light or pink-ish tan—probable undercure, unless artificially colored; dark tan to brown—good cure), by acetone extraction, water absorption, or degree of thickness recovery of the product after prolonged compression. Silicones are evaluated by surface (wetting) angle, and the other ingredients by specific quality and performance tests called out in their manufacturer's specifications.

The amount of binder present is a valuable control parameter and is determined by ignition at 1050°F of a dry, cured resin-glass sample and then calculating the percent weight loss (see Fig. 3-6).

Thickness and Density

These two parameters are so closely interrelated that, in the manu-facturing process, a change in one invariably produces a compensat-ing modification in the other. If a machine is producing at 1 in. thickness and 1 lb/cu ft density, and the thickness is doubled to 2 in., the density per inch of thickness would be halved. Hence, the quan-tity of fiber input to the machine must be doubled to maintain the product at 1 lb density. Since a near-uniform fiber production rate is desirable, the required gain in the fiber input per unit area is accomplished by halving the machine speed, thereby permitting twice as much fiber to accumulate.

In the manufacture of wool fiber, thickness is usually controlled by raising or lowering a set of "flights" or flat segmented elements on a chain drive which contact and compress the top surface. These move at the same speed as the bottom or collecting open-mesh conveyor. The flights are also constructed of an expanded metal or other open-mesh material to permit passage of heated air in the forced-draft curing oven.

Ultimate or specified thickness values of glass fiber and associated wool products are determined by the Gustin-Bacon "measurematic" null-balance device (see Fig. 3-7). In this unit the pressure of only a 3 g weight (to depress the few protruding surface fibers) is exerted by a plate which contacts the top of the test sample. Thicknesses vary in fiber glass end products from $\frac{1}{2}$ in. to as much as 8 in.

The accompanying density in blown fiber glass wool products is determined solely by weight of a sample 1 sq ft in area. Density may be made to vary from $\frac{1}{2}$ lb to as much as 7 lb/cu ft in some board

Fig. 3-7. Counterweighted null balance used to measure thickness of wool products as manufactured.

products. The upper limit on the flexible roll goods is approximately $2\frac{1}{2}$ lb/cu ft.

Hence it can be seen that many combinations of wool thickness and density are possible. Most product applications are based upon the best combination of the two to fulfill requirements of thermal, acoustical, or other service with performance balanced against cost. The close and necessary relationship between thickness and density will become more evident in the ensuing descriptions of individual products and their performance. (Fiber glass product density should not be confused with glass density mentioned earlier. Glass density refers to the factor of increase of the solid glass substance over the weight of an equivalent volume of water taken as unity.)

Percent Shot

As indicated, some of the processes generate a larger percentage of glassy beads or "shot" than others. The shot is often mobile, that is, not attached or adhered to adjoining fibers. Hence it may be removed by mechanical manipulation of a sample and weighed as a quality determination.

Percent Recovery

The degree of recovery after compression in insulation or wool products relates directly to the thickness which the manufacturer guarantees in his finished product specifications. The specifications for the product you want to purchase must be met under any and all conditions.

An austere condition exists in manufacture and packaging of either flat or roll-type insulation products. Unfortunately, they are usually compressed to conserve shipping space.

It would be most disconcerting to allow a 3 in. construction space for insulation, and when the material arrived for installation, find that it filled only a portion of the allotted space. In such an instance, naturally, the thermal efficiency and resistance to heat flow would be different than that originally designed for the building. Therefore, the industry sets and maintains rigid standards for recovery of the products to specified values.

The percent thickness recovery is influenced by the following: the original flight setting (usually original production thicknesses are slightly over specification); thickness itself (greater thicknesses generally have lower percent recovery); density (lower density-lower recovery); tightness of compression, rollup, etc., in packaging for shipment; type, age, formulation, and degree of cure of the bonding resin; and degree of relative humidity in the storage area (packaged insulation should be sealed inside non-moisture-transferring membranes).

Other Properties

Other functions of fiber glass and related mineral wool products such as resistance to heat transmission (thermal insulation), acoustical or sound absorption, "efficacy" as a filtration medium, and others will be detailed in the ensuing discussions of specific product applications and performances.

BUILDING INSULATION

Thermal Insulation—Homes

Insulation of homes against heat loss (winter) and heat gain (summer) probably represents the largest single usage for fiber glass and mineral wool products. Many different areas of the home may be thermally protected: ceilings, sidewalls, perimeters of slabs, floors, etc. Not only are many different types of available insulating materials used, but the way various components perform in combination must be taken into consideration in analyzing for the complete insulated structure, either in retrofitting or in new construction.

An understanding of the way insulation performs should start with consideration of the basic units of heat and related definitions.

HEAT LOSS DATA AND CALCULATIONS

(Reprinted from Johns–Manville Engineering Data on Building Products, #39Q, Feb. '75 or #FGBI-1, Nov. '75).

In the United States the basic unit of heat is generally considered to be the British Thermal Unit or, as it is commonly called, a Btu.

BRITISH THERMAL UNIT (Btu)—the amount of heat required to increase the temperature of one pound of water one deg Fahrenheit.

The basic unit of heat flow: i.e. the amount of heat that will be transmitted through a unit of material in a given time, is known as the thermal conductivity of a material.

THERMAL CONDUCTIVITY (k) = Btu per (hour) (square foot) (Fahrenheit degree per inch of thickness)–the amount of heat expressed in Btu that is transmitted through one square foot of a material *one inch thick* during a period of one hour when there is a difference of temperature of one Fahrenheit degree across the two surfaces of the material.

When thicknesses other than one inch are considered the term thermal conductance is used.

THERMAL CONDUCTANCE (C) = Btu per (hour) (square foot) (Fahrenheit degree)–the amount of heat expressed in Btu that is transmitted through one square foot of the material of a given thickness other than one inch during a period of one hour when there is a difference of temperature of one Fahrenheit degree across the two surfaces of the material.

When reference is made to a structure such as a wall which is composed of several different types of materials, air spaces, etc., the term overall coefficient of heat transmission is used. Unlike the k factor or the C factor where the heat flow is measured from surface to surface of a solid material, this unit also takes into consideration the insulating value of the air films formed on the surfaces of the materials.

OVERALL COEFFICIENT OF HEAT TRANSMISSION (U) = Btu per (hour) (square foot) (Fahrenheit degree temperature difference between air on the inside and air on the outside of a wall, floor, roof or ceiling)–the amount of heat expressed in Btu that is transmitted through one square foot of a structure during a period of one hour when there is a difference in temperature of one Fahrenheit degree from the air on one side of the structure to the air on the other side of the structure.

To facilitate calculation, the unit THERMAL RESISTANCE (R) may be used. Its value is equal to the reciprocal of the above heat transfer factors. $R = 1/k$, $1/C$, $1/U$. Thermal resistances may be added arithmetically to give the overall resistance of a structure. When calculating heat loss problems, it is well to keep in mind that heat always travels from hot to cold, i.e., from a higher temperature to a lower temperature and the temperature drop across a material is directly proportional to the thermal resistance of that material.

The use of the above information can best be illustrated by examples as follows:

Example: What is the thermal resistance (R) of a material that has a thermal conductivity (k) of 0.33 Btu in per hr per sq ft per °F?

Answer: $R = \dfrac{1}{k}$. Therefore, $R = \dfrac{1}{0.33} = 3.0$ per inch of material.

Example: What is the thermal resistance of 4 in. of this material?

Answer: $R = 4 \times 3.0 = 12.0$.

Example: What is the thermal conductance (C) of 4 in. of this material?

Answer: $C = \dfrac{1}{R} = \dfrac{1}{12.0} = 0.083$ Btu per hr per sq ft per °F.

Example: What is the overall coefficient of heat transmission (U) of an exterior frame wall composed of ½ in. gypsum board, 2 × 4 in. studs, medium spinsulation®, ½ in. weatherbrace sheathing®, and ½ × 8 in. bevel wood siding?

	Resistance
Answer: Air film resistance (still air)	0.68
½ in. Gypsum board	0.45
Medium Spinsulation® (insulation only)	7.
Air space	1.01
½ in. Weatherbrace Sheathing®	1.22
½ × 8 in. bevel wood siding	0.81
Air film resistance (15 mph)	0.17
Total Resistance	11.34

U Factor $= \dfrac{1}{R} = \dfrac{1}{11.34} = 0.088$ Btu/hr/sq ft/°F temperature difference

from the air on the inside surface to the air on the outside surface.

Hence, all materials used in home or other construction may be evaluated for thermal conductivity, or thermal conductance, and R or U factors determined. The thermal conductivity of an insulating material may be determined by the guarded hot plate, ASTM Method C177-45 (see Fig. 3-8) or the guarded hot box, ASTM Method C236-60.

In order to prevent a deleterious change in the ability of a fibrous insulating material to resist transfer of heat, it is necessary to apply a vapor barrier on the warm side. This membrane prevents moisture present in the room air from migrating into the insulating structure and condensing when the dew point is reached (through the inside-to-outside temperature gradient). Such an occurrence naturally lessens the R (resistance) value of the insulation and decreases its effectiveness by actually increasing its conductivity.

If unfaced insulation is used, there must be installed on the warm side a non-vapor-transmitting film, board, or any of the other vapor-resistant building materials. Vapor barrier membranes actually ad-

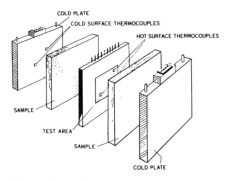

Fig. 3-8. Composite illustration showing a test bank of three low temperature guarded hot plates plus a schematic diagram of the operating components. The guarded hot plate is considered to be the most accurate technique for measuring thermal conductivity.

hered to the fiber glass and related thermal insulating products as sold include asphalt-impregnated kraft paper, aluminum foil-facing over kraft paper, and many others.

The R factors for fiber glass home insulation as it varies with thickness are shown in Fig. 3-9. Minimum recommended thicknesses of insulation for ceilings, sidewalls, and floors are indicated on the plot. As shown in the sample calculations, total insulation effectiveness for a given construction section is determined by adding the R values for all components.

Determination of proper R values for the amount of thermal insulation required for heat efficiency and fuel savings, and for protection

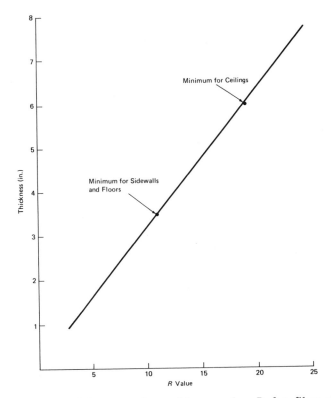

Fig. 3-9. Variations of heat-transfer resistance value R for fiber glass home insulation with change in thickness.

of air conditioning have been well worked out. The country has been divided into latitudinal zones based on prevailing average temperatures from North to South. Of course, more thermal protection is needed in zones farther North, and a higher R factor is stipulated for effective air conditioning for zones farther South.

Heating and fuel economy requirements are expressed as winter degree days (number of degrees of mean temperature for any day below 65°F and additive or cumulative for the entire winter season (6000 North to 2500 South)) and air conditioning requirements are expressed as summer cooling hours (all hours cumulatively during the cooling season that the temperature is 80°F or higher (400 North to 1500 South)). These zones have been detailed in the pertinent specifications listed at the end of this section.

It is certainly of interest to compare insulation effectiveness of fiber glass with other insulating materials. Table 3-3 lists R values for the normally usable thicknesses of the material.

After fiber forming, binder application, and cure, roll fiber glass for home insulation moves horizontally along a belt and the vapor barrier is applied using a resinous adhesive in small quantities. Machine widths may be up to 150 in., and windup speeds while compression is applied approach 10 ft/min. The vapor barrier film or paper is wide enough for fitting between standard stud spacings (16 or 24 in. centers) and the insulation is slit to accommodate the desired width.

Batting may also be produced off the machine. This material is in slabs of predetermined (cut off) lengths and stacked, not rolled. Compression is utilized in packaging.

Density of the type fiber glass for home insulation is in the range 0.3 to 0.6 lb/cu ft.

TABLE 3-3.   Comparison of R Values of Fiber Glass and Other Insulating and Construction Materials.

Material	Thickness (in.)	R value
Concrete	1	0.08
Common brick	4	0.80
Hardwoods—oak, maple	$3/4$	0.68
Plywood, flat siding	$3/8$	0.43
Softwoods, fir, pine	$3/4$	0.94
Sheathing	$25/32$	2.06
Vapor barrier film paper	—	0.12
Fiber glass roof insulation	$15/16$	3.70
Polyurethane roof insulation	1	6.67
Expanded polystyrene	1	4.00
Foamglas	$1\frac{1}{2}$	3.95
Air moving at 15 mph, nonreflective building materials	—	0.17
Still air, vertical with heat flow horizontal in nonreflective building materials	—	0.68
Mineral wool	—	—
Fiber glass rolls or batts	$3\frac{1}{2}$	11.0
Fiber glass rolls or batts	7	22.0
Fiber glass blowing wool	5	11.0
Double-glazed glass (windows)	$3/16$ in. air space	1.45
Fiber glass perimeter insulation	1	4.3
Thermal-acoustical fiber glass	4	11.0

Figure 3-10A shows a photo sequence of the major steps for installing home insulation between vertical studs. The utility flanges on the vapor barrier and their placement onto the studs are clearly shown (see also Figures 3-10B and 3-10C).

Fiber glass as used in this application is stable, may be made fire retardant, is verminproof, and will continue its function as a good insulator for the life of the dwelling.

Additional Products for Insulation of Buildings (Homes)

Several other adaptions of fiber glass insulations have been developed for homes and light commercial building construction. These are itemized here with brief descriptions of each.

Perimeter Insulation Slab or hardboard material 2 or 3 in. thick and 7 to 8 lb/cu ft density performs the function of preventing heat loss from the concrete floor base in basementless dwellings. There are several methods of construction to utilize perimeter insulation. However, in the most favorable method, the fiber glass extends vertically for a height equal to the thickness of the concrete slab, and also extends back under the slab for 12 or 14 in. all around the edge, and adjoining the vertical component. A vapor barrier is applied to the warm side between slab and insulation.

Blowing and Pouring Wool Loose pellets or large nodules of insulation are fabricated by picker action from cured fiber glass insulation. There are two types, No. 1 and No. 2, with the No. 2 being made from higher density material. The *R* values and corresponding thicknesses are presented in Table 3-4.

TABLE 3-4. Specifications for Fiber Glass Blowing Wool.

	NUMBER 1		NUMBER 2	
R Value	Minimum thickness (in.)	Minimum wt/sq ft (lb)	Minimum thickness (in.)	Minimum wt/sq ft (lb)
24	8½	0.52	7	0.69
19	6¾	0.41	5¾	0.56
13	4¾	0.28	3¾	0.36
9	3½	0.20	2¾	0.26

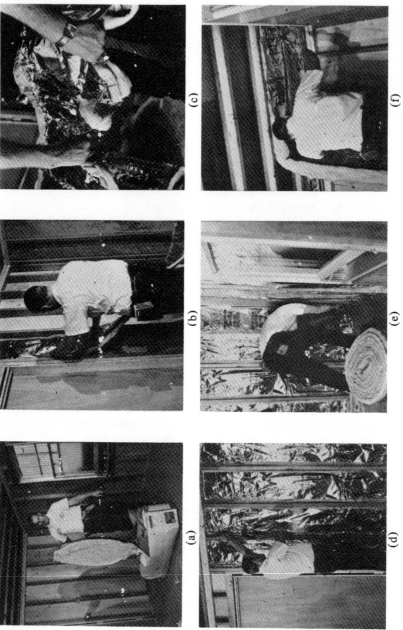

Fig. 3-10A. Sequence of steps in installation of home insulation. (a) Roll is removed from package. (b) Insulation is placed between studs as it is unrolled. (c) Most types are perforated at correct lengths for room height and may be easily separated by tearing, thus eliminating laborious cutting. (d) The vapor barrier paper is stapled in place along the inside stud facing. Cutting and fitting are employed to fill all shorter and narrower spaces between studs with the insulation. (e) The unrolled insulation strips are stapled at both bottom and top plates of each wall section. (f) Ceilings are also insulated by unrolling and pressing into place with one hand, stapling with the other. (*Courtesy Johns-Manville Sales Corp.*)

Fig.3-10B. Application of fiber glass blowing-wool insulation. (*Courtesy Owens-Corning Fiberglas Corp.*)

Fig. 3-10C. Application of home insulation in construction of a typical Japanese dwelling. (*Courtesy Nippon Glass Fibers Co., Ltd.*)

This type of macerated insulation may be either blown or poured in place as on flat areas (between joists in attics) or placed in sidewalls. The blowing equipment is standard gear well developed and suited for the purpose. Many fine points of installation technique are necessary in the procedure, and manufacturers' or distributors' instructions must be carefully followed. (See Fig. 3-10B.)

Sill-Sealer Insulation Strips 1 in. thick and usually 6 in. wide are provided in roll form for installation over the top of the masonry foundation and under the base framing sill plate. This insulation prevents air leaks and drafts from rising up into the dwelling, eliminates the need for caulking, and also keeps out insects, dirt, and dust.

Other Products and Adaptations Many other specialty products have been adapted to satisfy particular insulation requirements. These include rigid roofing board, masonry wall insulation batting (unfaced), reverse-flange insulation for use under floors in unheated or crawl spaces with stapling flange on the "breather" not the vapor barrier side, and thermal-type batting for acoustical use over ceiling tile or between room walls.

Specifications Home Insulation-Federal Specification HH-I-521E

Type I   = Friction-fit unfaced batting
Type II = Kraft-faced insulation
Type III = Foil-faced insulation

(Vapor transmission of barrier films also covered in these specifications.)

Blowing and Pouring Wool—Federal Specification HH-I-1030A Types I and II.

See also HUD-FHA Specification, "Minimum Property Standards for One and Two Family Dwellings," No. 4900.1 (November 22, 1974), and NMWIA interpretations of same; and "Insulation Manual, Homes and Apartments," prepared by NAHB Research Foundation, Inc.

Thermal Insulation–Metal Buildings

Enigmatically, at times rain never comes down outside when needed, and serious droughts occur. Then again, in the case of metal buildings, it may "rain" inside where it is never needed or desired. In the first instance, clouds may be "seeded" to induce rain. The second occurrence, caused by condensation due to the rapid temperature changes possible in metal buildings, is controllable by using proper insulation together with the best available vapor barriers so that condensation is avoided.

The first metal buildings of any import, the government and farm quonset huts, were difficult to insulate due to the absence of internal framing members. The quonsets could only be insulated by spray of cellulosics, plastic foam, or even macerated newspapers.

Improvements in the technology of mass fabrication of metal structural elements resulted not only in rapid economical construction, but also provided internal framing to which insulation could be applied.

Presently, 20% of all low-rise nonresidential structures are metal. The rate of new starts is double that for any other material used in this type construction, and is increasing at 10% per year.

The sudden incidence and rapid growth of this type structure have provided the material fiber glass with an excellent opportunity to solve many of the concomitant problems. In addition to the internal condensation problems discussed, metal buildings possess an extremely fast response to external heat (sun) or cold (winter) immediately transmitting these temperature changes to occupants or stored materials. Neither tilt-up slab concrete nor wood structures respond as rapidly. Also, the familiar sounds of "hail on a tin roof," or oil canning due to nonuniform thermal expansion or contraction causes audible noise pollution that may be readily treated by utilizing the acoustical properties of fiber glass.

Three main types of fiber glass insulation products have been developed to satisfy the requirements and resolve the problems posed in the construction and habitation of metal buildings.

Blanket Insulation

This is a roll-goods product available in various thicknesses and widths up to 72 in. It is laid across the top of the Z-shaped purlins (roof supports) or on the outside of the girts (sidewall bracing) before the outside building elements are added; a suitable vapor barrier membrane either decorative or only utilitarian is first installed on the warm side. Also, all metal roof sheets are sealed at each joint or lap.

TABLE 3-5.   Amounts of Insulation Recommended if Gas Heat is Used.*

Stripe	Approximate location in U.S.	Required blanket thickness (in.)	Corresponding R factor
No stripes	Northern Minnesota, Idaho, North Dakota	Over 4	16
No stripes	South to New York City, Peoria, Denver, Salt Lake City	4	13
1 stripe	South to Richmond, Tulsa, Sacramento	3	10
2 stripes	South to Columbus, Georgia; Vicksburg; San Angelo, Texas; Phoenix	2½	8
3 stripes	All South Central and Southeastern U.S.	2	6

*Note If gas heat and electric air conditioning are coinstalled, recommended R value is 13 for almost entire U.S.

The Thermal Insulation Manufacturers Association has issued standards PEB 202 and PEB 203 governing use of blanket metal building insulations. The country (U.S.) is divided into belts requiring varying insulation resistances if gas heat is used. These are correlated with selected stripe markings applied by the manufacturer to designate thickness, hence R value. The specifications should be consulted for detailed location of the area bands, but Table 3-5 shows a general coordination of these bands with recommended blanket thicknesses and R factors.

The blanket-type insulation does not generate any odor in the building, does not cause corrosion in contact with metals used (steel, copper, or aluminum), and carries an FHC 25/50 UL fire hazard classification. It also meets the requirements of Federal Specification HH-1-558B, Type I, Class 6.

Rigid Insulation Board

This material is a stiff, rigid fiber glass adhered to any of several facings and supplied in rolls up to 60 in. wide. It is unrolled at the site and placed across the purlins or girts under or behind the metal roof skin or exterior siding. Thicknesses vary from $1\frac{1}{2}$ in. ($R = 5$) to $2\frac{1}{2}$ in. ($R = 10$). The facings are either heavy vinyl (.004 in.), foil laminated with fiber glass and kraft paper, or other heavier puncture-resistant material. These facings are not only good vapor barriers, but are adaptable in the major uses of the rigid board insulation for the inside of tennis courts, buildings for light manufacturing or warehousing, etc., where mechanical impact or abuse is likely to occur.

Engineered Systems for Increased Thermal Performance

These systems are newcomers developed to guard further against excessive energy loss from metal buildings, to reduce stray heat losses such as direct heat transmission through unguarded purlins, etc., and to provide an easily installed medium for reinsulation of existing buildings, improving appearance and servicability.

The system can be either added to a metal building already containing either the batting or rigid roll insulations described in the foregoing, or included in new construction. It consists of a $5\frac{1}{2}$ in. thick unfaced fiber glass batting mounted between the purlins or behind girts, and covered with a $1\frac{1}{2}$ in. higher-density board bearing a heavy vinyl facing for both appearance and impact resistance factors.

The third element is a C-shaped extrusion that is installed on the bottom of the purlins, and holds a vinyl-faced fiber glass cap strip at least $1\frac{1}{2}$ in. thick. These components greatly reduce heat loss by conductivity through the purlins.

When installed with the standard $2\frac{1}{2}$ or 3 in. roof insulation, this system will provide an increase in the R factor to 20, as determined for the composite structure using the guarded hot box. Figure 3-11A shows an installation of this system in a new indoor tennis court construction. Installation in the sidewalls is shown in the main photograph, while in the insert, the cameraman has looked up to view the partially filled purlins and channels (before and after installation).

As regards profitability and savings, it may be stated that economic thickness of insulation for pre-engineered metal buildings keeps total owning and operating cost as low as possible, taking into account the cost of fuel and electricity for heating and cooling, the cost of the insulation itself, and other factors. As the amount of insulation in the roof and the walls of a building is increased, energy consumption declines. The thickness of insulation that produces the lowest annual sum of the costs of energy, equipment, and insulation is the economic thickness. This is shown graphically in Fig. 3-11B.

Fig. 3-11A. Application of a completely engineered system of fiber glass insulation to the interior of a new metal building, an indoor tennis club. The main photo shows installation of the vinyl covered material on the sidewalls, and the insert photo shows the material partially in place between and on the face of the purlins. The finished interior is represented at the top of the page opposite. (*Courtesy Johns-Manville Sales Corp.*)

Fig. 3-11A. (*Continued*).

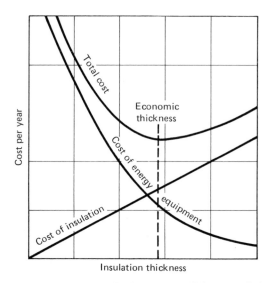

Fig. 3-11B. Means of determining the lowest possible sum of the annual cost of energy and equipment and the annual cost of insulation is shown graphically. As insulation thickness is increased its cost increases, but not proportionally because there is little added labor cost. With more insulation, there is less heat loss; less equipment capacity may be necessary, thereby reducing the consumption of energy—and its cost. The lowest combination of these three factors—*energy, equipment, insulation*—indicates the economic insulation thickness, and the point where the most money can be saved.

Insulation of Mobile Homes, Recreational Vehicles, and Packaged Housing

The Federal Department of Housing and Urban Development has issued construction and safety standards for all mobile home designs, and requires approval based on energy requirements. The Specification, No. 40 FR52706, became effective mid-1976. It requires that all walls, ceilings, and floor cavities be insulated.

In this specification, the United States territorial areas are divided into 3 zones; hot to cold they are (I) All southern U.S. latitudinally through the northern border of Oklahoma and including California; (II) all northern U.S.; and (III) Alaska.

In addition to the insulation requirements, storm windows or insulating glass must be included on mobile homes in Zones II and III. Also, each design must provide for condensation control, guard against air infiltration through walls and accompanying heat loss, and must possess low heat-transmission U values for use in each Zone as follows: $I = 0.157\ U$; II $= 0.126\ U$; and III $= 0.104\ U$.[3]

Maximum R values have not been set. However, those extant are ceilings—19, floors and walls—11. One accurate assessment of the amount of insulation and its effectiveness is the corresponding quantity of visual window area. Of course, for larger amounts of insulation installed the permissible amount of window area increases. The allowable window area becomes reduced, however, with increase in the trailer floor area from 12×50 ft through 14×63 ft to 24×70 ft.

Basic construction of mobile homes is wood framing. However, one fabricator of larger-sized manufactured housing units (components erectable at the site) has employed steel girders and framing to provide better dimensional stability and freedom from warping. No problems are created with placement of the fiber glass insulation in the metal framing.

Fiber glass duct board fabricated at mobile home assembly sites into conduit for heating and air conditioning has more than proven its worth. (See Pipe and Duct Insulation.) Flexible ducting is also employed. The composite illustration, Fig. 3-11C, shows installation of vapor-barrier-faced insulation in sidewall studding, laying a 12-ft wide blanket for floor insulation, and placement of a fabricated rectangular fiber glass duct inside the roof-truss framing of a mobile home.

Fig. 3-11C. Composite illustration showing insulation of mobile homes during construction: (left) installation of vapor-barrier-faced batting in sidewall studding, (upper right) incorporation of a 12 ft wide blanket of floor insulation, (lower right) placement of a rectangular fabricated all-fiber glass duct inside roof-truss framing. (*Courtesy Johns-Manville Sales Corp.*)

ACOUSTICAL INSULATION FOR BUILDINGS

Introduction, Dimensions, and Testing

To resolve the long-disputed point questioning whether noise exists if there are no detection devices present, "sound" refers both to the occurrence itself and also to the human or mechanical facilities necessary to sense it.   Sound results when some vibrating medium compresses and decompresses the surrounding air.   This may result in spherical emanation of sound waves if the vibration is sustained, or in only a single shock wave for one mechanical excitation.   The degree of compression and accompanying decompression is measured as loudness level on a "decibel" scale (dB), ranging from 0 to 120 or more, or as sound pressure in microbars (range 0.0002 to 200) for the human hearing range.   Table 3-6 lists characteristically recog-

TABLE 3-6.   Recognizable Sounds and Corresponding Noise Levels.

Example or source	Loudness level (dB)	Sound pressure (microbars)
Lower threshold of audibility	0	0.0002
Sound of rustling leaves	10/20	/0.002
Whisper at 5 ft distance	20/30	
Quiet conversation	30	
Quiet radio	40	0.02
Normal office background noise	35/50	
Average conversation	50/60	/0.2
Window air conditioner	55	
Typewriting pool, 9 machines	65	
Average factory	70	
Loud radio	80	2
Threshold of human hearing damage if sound is sustained	85	
Noisy factory	90	
Heavy city traffic	92	
Home lawnmower	98	
Auto horn at 23 ft	100	20
Riveter	110	
Jet aircraft at 500 ft altitude	115	
Threshold of auditory pain	120	200

nizable noises together with corresponding sound pressure levels. Loudness depends upon sound pressure and diminishes with distance from the source. The unit of loudness is the Sone with unity equal to 40 dB above the listener's lower threshold of hearing at 1000 Hz frequency.

Other important considerations in sound are velocity, frequency, and wavelength. Sound velocity in air is constant for all sound frequencies at 1120 ft/sec.[*] Audible frequencies vary from 20 to 20,000 cycles per second (Hz) with those for most everyday recognizable sounds ranging from 20 to 4000, the upper practical level of normal human hearing. Examples are the human voice—100 to 4000 cycles, and the piano—27 to 4186 cycles. Those frequencies used for acoustical measurement are the octave bands 125, 250, 500, 1000, 2000, and 4000. Sometimes the $\frac{1}{3}$ octave bands such as 100, 125, 160, 200, 315, etc., are added. The fact that sound velocity is a constant makes frequency and wavelength inverse, since wavelength is deter-

*At standard conditions of temperature and pressure.

mined by dividing velocity by the frequency. Hence, wavelengths vary from 56 ft at 20 Hz through 0.28 ft at 4000 Hz to 0.0056 ft at 20,000 Hz.

Quality of sound depends upon the vibrating source and contains more unknowns than knowns. A well-played violin is preferred to a hammer-on-anvil noise, and both are the result of an exciting or vibrating source causing systematic, successive air compressions and decompressions (waves). Also, combinations of sounds are a factor, but this is getting into the learning or personal preference realms.

Another factor that accounts for differences in apparent sound sensation to the ear is overtones. These are the result of a vibrating source producing several different frequencies at the same time, and this leads to the major considerations necessary in sound deadening, absorption, or insulation.

There are four areas where treatment of sound for acoustical improvement may be carried out.

1. At the source, with use of quiet appliances, machinery, plumbing fixtures, resilient mounts, and use of sound-absorbing baffles or enclosures to contain the noise.
2. Reduction of transmission of sound through walls, floors, and ceilings.
3. Quieting by placement of sound-absorbing materials to reduce reverberation and noise levels within a room.
4. Acoustical correction by design, which includes shape of room or auditorium space plus proper amounts of sound-absorbing materials applied to walls and ceiling surfaces.

Fortuitously, the material fiber glass offers the technical capability and product adaptability to serve as sound reducer and absorber for almost every facet of each of the above classifications. Sound absorption increases with thickness of the fiber glass blanket or part, and with reduction in fiber diameter. Variations in batt density have a significant effect on acoustical properties, but not as marked an effect as batt thickness and fiber diameter (see Table 3-7). For cosmetic purposes, a thin membranelike material or paint may be used to cover fiber glass acoustical material which is to be directly exposed to the vibrating sound source or the air between. This membrane material itself vibrates, thus transmitting the sound energy into the body of the fiber mass where it is dissipated.

TABLE 3-7. Effect of Fiber Glass Batt Filament Diameter, Thickness, and Density on Acoustical Absorption.

Filament diameter (μ)	THICKNESS = 1½ in. Density in lb/cu ft required to provide acoustical absorption of NRC = 0.65	THICKNESS = 2 in. Density in lb/cu ft required to provide acoustical absorption of NRC = 0.75
6	4.80	3.9
3	1.78	1.45
2	—	0.81
1	0.37	0.30

Two major test methods have been designed for evaluation of the effectiveness of all acoustical insulations including fiber glass.

The first, ASTM Method E 90-61 T (which see) is a laboratory method of measuring sound transmission loss through walls and doors. It is also adapted to evaluation of ceilings and wall components such as office partitions. The units are decibel numbers for a sound transmission class (STC) selectable based on the type or level of noise it is desired to quell. For office partitions or ceilings, it is only required to mask speech or minor office noises, but for homes or apartments, appliances, children, etc., must be masked. Hence, a sliding or graded scale of sound transmission loss in decibels versus frequency was predetermined, and the actual performance curve for a fiber glass or other acoustical material is related to the standard curve and is adjusted correspondingly. The higher the STC range, the better is the resistance to sound transmission.

The second test method, determination of sound reduction by impingement onto and reflection from the facing of an acoustical material, may be carried out by two separate test procedures.

These are the acoustical impedance-tube test, ASTM C384-58 (see Fig. 3-12), and the reverberation room, ASTM C423-66 (see Fig. 3-13A). The sound incidence is normal (perpendicular) to the sample in the impedance tube test, and random (0 to 60°) in the reverberation chamber. Both methods may be used to determine the fraction of sound absorbed by varying the frequency through the range 125 to 4000 Hz, and taking measurements at set points in between. From these data, the noise reduction coefficient may be determined. The NRC is the arithmetic average of the individual

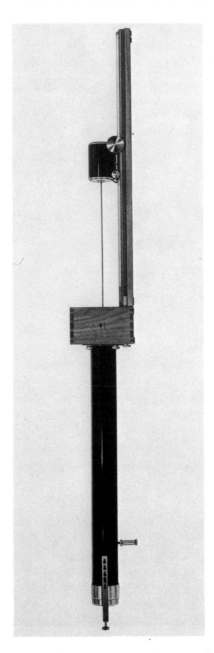

Fig. 3-12.   Acoustical impedance test apparatus.   Diameters of the test units may be either $1\frac{1}{8}$ or $3\frac{7}{8}$ in. I.D. The test sample is mounted at one end and the wavelength of sound varied by positioning the probe.   (*Courtesy R & K Instrument Co.*)

Fig. 3-13A. Reverberation-chamber test room for measuring sound absorption of fiber glass and other acoustical materials. Purpose of the test room is to build up the sound pressure level and preserve as much echo as possible, evaluating the property of the test material in absorbing the sound generated. This chamber has a volume of approximately 11,250 cu ft. The flat-surfaced wall, ceiling, and floor are reinforced concrete. The chamber has its own founda-

Fig. 3-13B.   Close-up view of wall structure of anechoic acoustical test chamber. The tapered wedges are fabricated from fiber glass insulating material. (*Courtesy Owens-Corning Fiberglas Corp.*)

sound absorption coefficient determined at each frequency level. The NRC varies between 0.0 and 1.0 with the higher numbers indicating a greater sound-absorbing ability for the material.

In contradistinction to the reverberation chamber, and not to be mistakenly identified with it, is the quiet-room or anechoic test chamber.   In this unit the walls, floor, and ceiling are completely surfaced with inwardly protruding, hyperbolically tapered sound-absorbing wedges, also usually fabricated using fiber glass.   (see Fig. 3-13B.)

The function of the reverberation chamber is to reflect sound and make it persist in order to evaluate the ability of the test medium

---

tion, separated from that of the surrounding building. The large vanes at the top of the room rotate slowly to insure diffuse sound within the chamber. The corrugated plastic stationary panels also assist in diffusing the sound. The usual test specimen is 72 sq ft (8 × 9 ft), and may be mounted in any one of several standard fixture types depending upon typical end-use conditions. (*Courtesy Johns-Manville Corp.*)

to absorb it. However, the purpose of the quiet-room is to measure the sound generated by a motor, appliance, transformer in a light fixture, or the amount of sound reflected from a rigid appliance surface.

In the reverberation room, reverberation time is defined as the number of seconds required for the sound pressure level to diminish 60 dB below its original level at the selected frequency. For the anechoic chamber, the sound emanating from a source or being reflected from a surface is measured directly in decibels and compared to levels of human tolerance.

Other important properties to be determined for fiber glass and other acoustical materials are physical strengths, resistance to ambient air flow, light reflectance, fire safety rating, and more specific parameters to be indicated later.

Three separate groups of acoustical fiber glass products will be discussed: (1) structural internal noise suppressants for walls, etc., (2) residential and commercial ceiling components, and (3) items available for custom commercial and industrial applications and their usage.

The acoustical industry is fairly young, having had its first practical application problems about 1895. Matted animal felt and macerated, bonded cane fibers formed the first sound-absorbing materials. Fiber glass and allied fibrous materials are playing an interesting role in alleviating the difficult and sophisticated acoustical conditions that present themselves today.

Thermal-Acoustical Batting

The line of demarcation between beneficial thermal and acoustical performance is not well defined. It is sufficient to say that, if fiber glass material is installed to perform one of the functions, a special bonus is usually gained, because the other function will be fulfilled also. There are three such major applications for fiber glass specific to the building industry, and briefly described as follows.

Fiber Glass in Wall Construction for Reduced Sound Transmission

As stated, sound does not travel through a wall, but successive sound waves impacting a surface set up a vibration or diaphragmlike action, and the wall elements become individual sound-producing sources. Many different structural factors bear on this condition, and the sound may not possess the exact characteristics as the original vibrating source. However, if the original source is noxious, the secondary source will be also.

Hence, in home or building construction, any method of creating a discontinuity or lack of direct mechanical contact between wall elements, accompanied by filling the internal spacing with fiber glass insulation, will contribute substantially to noise reduction. The several systems of improved wall construction (between rooms or on outer walls) are summarized in Table 3-8. Sound-reducing effectiveness is indicated by the STC values.

Thermal-Acoustical Insulation for Improvement of Existing Construction

So-called prepanel insulation coverings for existing masonry walls, and others, constitute methods of effecting improvements in acoustical, and concomitantly, thermal performance. Prepanel insulation is a light-density, $1\frac{1}{8}$ in. thick material placed between $\frac{3}{4}$ in. furring strips over masonry or older walls in reconstruction or remodeling. When the decorative interior paneling is applied, the insulation is compressed to the $\frac{3}{4}$ in. furring-strip thickness. In this function, in addition to providing thermal and acoustical protection, the insulation takes up any variations in the furring-strip thicknesses, adequately fills the void, and relieves the hollow, drumlike sounds which normally occur in unsupported paneling.

Thicker batting is available for placement over masonry walls. The R values are 2.9 and 5, respectively, for $\frac{3}{4}$ and $1\frac{1}{2}$ in. thick prepanel insulation. A vapor barrier is usually necessary on the warm side. Acoustical function of the insulation may also be impaired by presence of moisture in the fibrous structure.

Additional Insulation for Acoustical Ceilings

Suspended ceilings (discussed in the next section) sometimes require additional insulation to bolster both thermal and acoustical properties. Batts in 2 X 4 ft cut shapes with unfaced, nonoverlapping kraft or foil-faced surface treatments, or roll stock are supplied. These batts are laid in place over the suspended ceiling tile. Whereas a

**TABLE 3-8. Insulation and Wall Constructions Contributing
to Reduced Sound Transmissions.**

Description of construction condition	Quantity of fiber glass insulation	Corresponding STC value
1. Standard wood 2 × 4 in. studs	None	31
2. Double stud wall (2 rows, 2 × 4 in. studs on 2 × 4 in. plate separated by 1 in. with ⁵⁄₈ in. gypsum dry wall)	None	43
Same	3¹⁄₂ in. thick	55
Same	3¹⁄₂ in.—2 layers	59
3. Staggered wood studs (staggered rows of 2 × 4 in. studs on 2 × 6 in. plate with ⁵⁄₈ in. gypsum board)	None	42
Same	3¹⁄₂ in. thick	49
Same	3¹⁄₂ in.—2 layers	52
4. Single studs with resilient channel (2 × 4 in. studs on 2 × 4 in. plate with gap created by sound reduction channel and ⁵⁄₈ in. gypsum board)	None	40
Same	3¹⁄₂ in. thick	51
5. Steel-stud walls—3⁵⁄₈ in. steel studs mounted 24 in. centers on steel mounting track with ¹⁄₂ in. gypsum wall board.	None	45
Same	3¹⁄₂ in. thick	49
Same	3¹⁄₂ in. thick with double layers of gypsum wall board both sides	56

Note 1: Other methods of reducing sound transmission through walls: use of wall facings of different thicknesses to avoid sympathetic resonance or vibration; use of caulking at wall plates (top and bottom) and both sides; tape and seal all holes in the gypsum board; avoid back-to-back electrical outlets or fixtures, connecting ducts, or any direct transmitter of sound vibrations from one surface to the other. Vapor barrier should be used when required.
Note 2. Fire resistance rating for most of the above is 1 hr. Fire resistance rating may be improved by installing three or four thicknesses of insulation instead of two.

single suspended $1\frac{1}{2}$ in. acoustical ceiling panel will possess an NRC of 0.80, over-addition of a $3\frac{7}{8}$ in. thick acoustical batt will increase the NRC to 0.90. Thermal R values for $3\frac{1}{2}$ and $6\frac{1}{2}$ in. thick batts alone are respectively 11.0 and 22.0. When placed over representative suspended ceiling panels, the R values increase to 14.1 and 25.1, respectively.

The added weight of acoustical batting may cause distortion or sagging of the ceiling panels. Limitations are: no added batting for panels $\frac{5}{8}$ in. thick; limit of $3\frac{1}{2}$ in. thick batts on $\frac{3}{4}$ in. panels; and 6 in. maximum batting thickness on 1 in. panels.

Acoustical Ceiling Materials

With the advent of greater awareness by building engineers and architects and now government agencies, for the protection of the senses, particularly hearing, together with other safety measures, the use of fiber glass and acoustical types of fibrous insulators in ceilings has really flourished. Ceilings are the first line of acoustical defense in buildings because they present the largest uninterrupted surface or minisurfaces for sound reflection. Unless properly treated, reverberations magnify normal sounds to the point of complete interference and unintelligibility, and if sustained, prematurely induce fatigue or damage hearing.

The typical characteristics of fiber glass ceilings for improvement of acoustics in homes, schools, offices, hospitals, and industrial buildings are presented herewith. Substantiating performance data are included.

Materials

Fibrous materials used are fiber glass and other inorganic mineral fibers. The fiber glass is made by processes already described. The mineral-fiber acoustical materials are first produced by the blown attenuation processes, and the acoustical products per se are felted from a wet slurry. Binders are applied in both cases to preserve product integrity. The fiber glass materials are slightly less dense than items produced from mineral fibers. The advantages of each are well documented; also several different beneficial properties and characteristics may be combined in the same acoustical ceiling component, as will become evident in the following.

Dimensions and Suspending Systems

Thicknesses of acoustical ceiling materials vary between $\frac{1}{8}$ and 2 in. The thinnest, $\frac{1}{8}$ in., comprise the class of fiber glass sculptured panels, manufactured in a manner similar to that described for automotive topliners, hoodliners, etc. (see Automotive Insulations) and are illustrated in Fig. 3-14A, Parts A and B. Mineral fiber acoustical elements are generally fabricated to $\frac{1}{2}$ or $\frac{5}{8}$ in. thicknesses. The larger, flat fiber glass components are produced in thicknesses from $\frac{5}{8}$ to 2 in. (Fig. 3-14A, Part C.)

As regards area, or length-width dimensions, the smallest size, designated as "tiles," are 12 × 12 in. and may go to 12 × 24 in., or 12 × 36 in. Next are "panels" which extend from 24 × 24 in. to 60 × 60 in. "Boards" overlap in the smaller sizes with panels, but may extend to 4 ft × 8, 10, or 12 ft as standard production, or even 4 × 16 ft if required and specially fabricated.

A

Fig. 3-14A Parts A and B. Two types of sculptured (molded) fiber glass acoustical ceiling panels; the first (dome shaped) installed in the ceiling of a swimming pool building, and the second (angular or "prismatic") installed in the ceiling of a large shopping mall. (*Courtesy Johns-Manville Sales Corp.*)

B

Fig. 3-14A.   Parts A and B.   (*Continued*)

Suspending ceilings for tiles may consist of either cementing to a plain, flat ceiling surface (for new construction) or to an existing ceiling (for remodeling).   Flush furring may also be placed on a new or existing ceiling and the tiles nailed or stapled to the strips.   Tiles are edged by kerfing or applying tongue-and-groove laps for proper fit and appearance.   The exposed edges of the tiles are usually bevelled 45°. Panels are mounted by laying into the tracks of upside-down metal tee-bars suspended from the ceiling with wood strips or wires.   The panels are usually square-cut 90° at the edge since they must rest on and be retained by a surface of the tee-bar $\frac{1}{2}$ in. or less. Some mineral fiber panels are rabbeted so that they are supported

C

Fig. 3-14A, Part C.   Flat panel, film-faced-type acoustical panels mounted in appropriately designed inverted tee-shaped track and installed in the ceiling of a modern engineering office.   (*Courtesy Johns-Manville Sud Americana, LTDA, Buenos Aires, Argentina, Subsidiary of Johns-Manville Sales Corp.*)

by the metal tee-bar but also extend below it, enhancing appearance. Tracks may be suspended a minimum of 3 in. from the existing ceiling.   Some suspension tracks are not visible from the floor level.

Larger panels and beams may be supported by tracks with a larger face, or by simulated beams affixed to vertical studding, the latter being a portion of the dwelling structure.   Panels and beams are usually square-cut at the edges.

With the suspension grid system, laid-in panels may be removed easily from the grids or tracks for cleaning.   Further, an intricate but workable system has been designed around one of the mineral-fiber acoustical panel types.   To permit access to the plenum space above the false ceiling for the purpose of servicing wiring, piping, etc., upper and lower matching angles are provided.   The lower matching angle is mounted in a horizontal slot in the side of the stationary mineral fiber panel, and the panels rest on a cross-tee between two main tees. The upper matching angle is likewise mounted in a slot in the side of the panel to be lifted, and rests over the top of the angle in the stationary panel. By applying light pressure from below, the two panels containing the upper matching angles may be made to operate like a trap door.

Dimensional stability of both mineral fiber and fiber glass acoustical ceiling components is excellent, with mineral fiber being somewhat superior because of its higher density and rigidity. One manufacturer of fiber glass panels provides a 10-year guarantee of the suspended elements against warping, buckling, or sagging. Another admits that sagging may occur if thermal-acoustical batting is laid over the suspended acoustical panels.

Aesthetic Appearance: Facings, Configurations, Contours

All tiles, panels, and boards, both fiber glass and mineral fiber, are created with the utmost consideration given to appearance. The reasons for this are obvious, since the acoustical elements are continually seen, must provide interesting patterns and blend with other room or area decor, in addition to being technically functional.

Mineral fiber tiles and panels may be supplied with textured surfaces. These consist of uni- or multi-directional pressure-induced fissures, pressed mottled surfaces, punched holes of various sizes and spacings, and combinations of all the above. All panels are painted with a thin coating of polymeric material which serves to decorate as well as permit transmission of the sound waves into the absorbing medium (see Figs. 3-14B and 3-14C).

Also, due to their integrity, density, and dimensional stability, the mineral fiber tiles and panels may be covered with either a perforated or unperforated aluminum membrane. The end products are used where cleanliness and freedom from dusting are required, as in hospitals and other medical facilities, food preparation areas, locker rooms, etc. These panels also possess special built-in membranes or cores to resist penetration and transmission of airborne moisture.

Finally, mineral fiber panels may be completely through-perforated. The application is, naturally, air conditioning of the ceilinged area or room, with the cool-air ducting located in the plenum above the acoustical panels.

Fiber glass acoustical components do not carry texturing in the body of the material per se, but may be painted, and also may carry

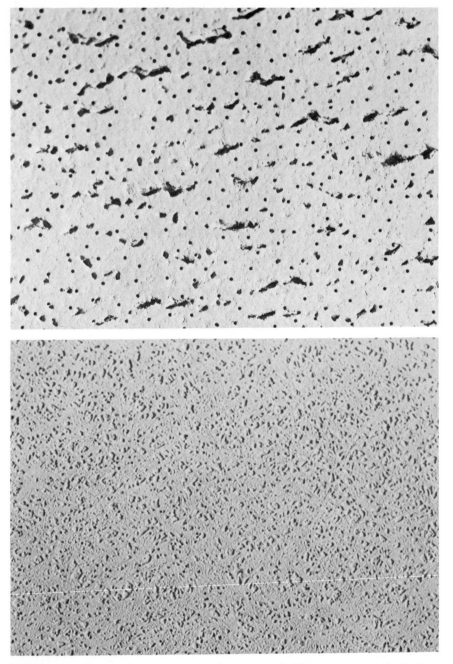

Fig. 3-14 B. Characteristic patterns for mineral fiber-type acoustical tiles.
(*Courtesy Johns-Manville Sales Corp.*)

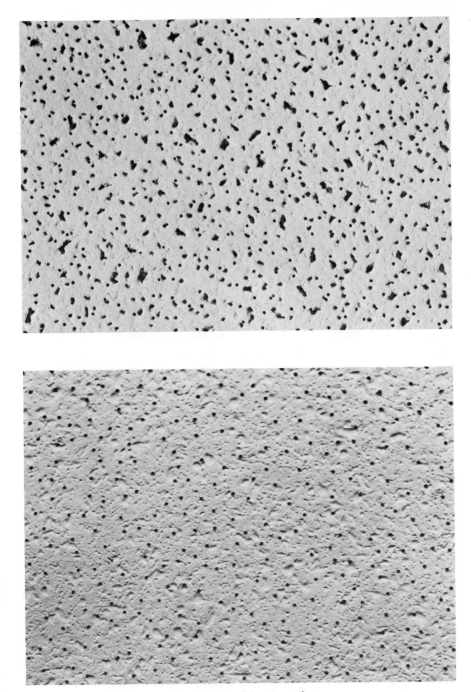

Fig. 3-14B. (*Continued*)

Fig. 3-14C. Above. Sprayed-on mineral fiber acoustical ceiling composition in Ontario Science and Technology Building, Toronto, Canada. Next page. Same composition in PATH Railroad Station, Jersey City, N.J. This material is completely inorganic and furnishes excellent fire resistance. (*Courtesy United States Mineral Products Co., Inc.*)

a popular, well-designed domed or sculptured configuration (see Fig. 3-14A). The remaining types of fiber glass acoustical panels and boards are all flat and their surface treatments are accomplished by application of either a thin, embossed polyvinyl chloride or similar type film, or by a decorative fiber glass fabric.

Light Reflectance

Illumination of an object varies with the square of the distance from the source, assuming normal (perpendicular) incidence. Obtaining optimum illumination of and light reflection from a fiber glass (or associated material) ceiling is further complicated by the fact that the illuminating sources do not impinge normally, but usually at a low angle. Light from a window strikes the ceiling surface at a

Fig. 3-14C. (*Continued*)

glancing angle.   Of the three types of lighting fixtures used with
suspended fiber glass ceilings, recessed, flush mounted, and pendant,
only the latter casts any appreciable quantity of light directly onto
the ceiling surface.   Further complications are introduced by interception of low-angle incident light by any downward-projecting
objects, i.e., the tile or panel edges in certain instances.   Manufac-

turers apply a bright but matte-surface white to gain maximum reflectance.

Through many design cycles and intensive consideration of customers' requirements, the result has been placement of illuminating sources so that the most uniform illumination is possible across the entire face of the ceiling. Notable deviations have been attempts to create interesting repetitive patterns of light and shadow using the contoured or sculptured fiber glass panels, or rabbeted-edge mineral fiber panels. Also the many and varied patterns and surface treatments available have shown that panel manufacturers have realized that "beauty is in the eyes of the beholder," and that they must provide many patterns and textures to satisfy the naturally wide variety of tastes.

Standards for determining the illumination of acoustical ceiling components are: (1) ASTM C523-68, in which light incident at an angle between 25 and 55° from normal is reflected back 75% or greater (adopted manufacturers' standard), and (2) Federal Specification SS-S-118A in which ratings LR-1 or LR-2 may be assigned to most all fiber glass and related acoustical materials.

Acoustical Ratings

Fiber glass acoustical panels and boards in the usual 1, $1\frac{1}{2}$, or 2 in. thicknesses, 1 lb/cu ft density, have higher noise reduction coefficients (NRC) from 0.70 to 0.95. The thinner molded "sculptured" panels test slightly lower at 0.50 to 0.75. Elements made from mineral fiber test slightly lower for NRC at 0.45 to 0.75, thus indicating that panels made solely from fiber glass have slightly superior capabilities for nonreflection of incident sound.

However, as regards sound transmission through the panels or other components, the mineral fiber materials are somewhat superior, showing high STC decibel retardancy from 35 to 50. It is possible to incorporate selective core materials within the surface boundaries of mineral fiber tiles and panels to further retard the transmission of sound.

This indicates that more sound is transmitted through fiber glass ceiling panels. This condition does not indicate an entire loss or failure. It actually favors installation of sound speakers above the plenum in an office, permitting muffled music sounds to traverse through the ceiling panels and mask office noises, or permit more private conversations by making nearby speech unintelligible.

Thermal Properties of Ceiling Components

Actual thermal insulation R values of mineral fiber tiles and panels $\frac{5}{8}$ to $\frac{7}{8}$ in. thick range between 0.51 and 2.43 (special core material included to gain the latter value). The fiber glass panels provide R values up to 4.08 for 1 in. thickness and 6.12 for $1\frac{1}{2}$ in. thickness. Thermal resistance R for the $\frac{1}{8}$ in. thick compressed sculptured fiber glass panels is poor at 0.51. As stated, thermal efficiency is improved by incorporation of lay-in thermal batting above fiber glass acoustical panels. Generally, however, in building design, thermal insulation is given fuller consideration including the acoustical elements, but also taking into account the interior-exterior building interfaces.

Ratings for Resistance to Flame Spread and Endurance Against Fire Transmission and Flame Penetration

Two systems are involved here for evaluation of acoustical materials. (1) The UL Tunnel Test for flame spread described in ASTM Standard E84. Both mineral fiber and fiber glass elements possess class 25 noncombustible ratings by this test. (2) The fire endurance test (ASTM E119) in which the ability of an acoustical material to delay heat transmission and flame penetration is evaluated. Tiles and panels made using mineral fibers show acceptable performance for this parameter. Time ratings from 1 to $2\frac{1}{2}$ hr are achievable, and these only after special proprietary treatments or formulations used in the manufacturing processes.

Systems for Custom Acoustical Installations

With the advanced technology gained from analysis of critical modern acoustical problems, and the products resulting from the accompanying research, manufacturers now have available a complete "stable" of products. Hence architects, designers, and acoustical engineers, when consulted, can draw upon this wealth of information and wide range of materials in solving specific problems.

Integrated Systems

In addition to the acoustical materials, entire coordinated systems have been developed and are offered as a package. Included are lighting fixtures that fit the modules planned for the size of the acoustical tiles, panels, or boards, and air conditioning or distribution components. The latter may be in the form of the perforated acoustical ceiling previously described, or may be composed of standard acoustical ceiling tile or panels adjacent to a hidden fiber glass or other duct extending down the center of the room and delivering conditioned air through a louvered opening or slit into the room. This system will be considered further under Pipe and Duct Insulation.

Another fiber glass acoustical insulating component that enjoys wide use and lends flexibility to the designer is surface wall panels. These come in various densities up to 6 lb/cu ft, thicknesses up to 6 in., and various facings, the most utilitarian of which is fiber glass cloth. These external wall treatments are useful in offices or larger meeting rooms when it is required to provide sound reduction over that of which the original ceiling treatment is capable.

Many installations, combining wall and ceiling panels, have been completed by acoustical contractors, jobbers, or distributors, in which areas already constructed have been rescued from almost complete desertion due to noise problems. Complete surface covering of ceilings and walls by faced acoustical paneling is not necessary, but judicious use of fiber glass panels based on knowledge and experience has resulted in the revitalizing of these rooms and areas for their originally intended uses.

The Open Office

So much planning, effort, and technology has been brought to bear on solving acoustical problems related to copiously populated offices that a special discussion is justified. With larger staffs required and energy critical, the move to the "open office" has begun in earnest. One aggressive and growing manufacturing firm selected the open-office concept for its new corporate headquarters because of the low cost of 30 cents per sq ft to move one office occupant versus $15 per sq ft moving from or into a private cubicle. Other sages of the industry predict that in 10 years almost 70% of all offices will be built around the open plan.

The stated advantages of the open office are more functional inter-personal communications, lower placement cost and greater adapta-bility to change, and more efficient use of space. The noise problems to which every denizen of such an office workers' landscaped "Garden of Eden" is subjected have been foreseen, measured, and handled in a most interesting fashion. The essential goal is to achieve speech privacy, while keeping total noise down to an acceptable level. Three major components are required to absorb office noises (typewriters, air conditioners, etc.) and make normal communicative speech private and unintelligible a distance of 3 to 4 ft away (only hearing one word in five). There are: (1) an acoustically efficient ceiling with as high an NCR as economically feasible, and accompanying acoustical wall treatment if necessary; (2) an electronic sound system of speakers installed in the plenum above the ceiling, tile, panels, or boards, emitting a uniform, steady low-level (NC = 40) noise (not music) which masks a major portion of the speech in the office below, and renders it unintelligible; and (3) a system of combined physical partitioning and acoustical screening panels separating but not necessarily surrounding each office module. These partitioning elements are essentially composed of a fiber glass core with internal metal septum and external decorative dacron polyester fabric. Figure 3-15 illustrates use of these sound-screen panels in an open office area, and an accompanying inset view diagrams their internal construction.

Complete acoustical technology, too detailed for inclusion here, has been published.[4] Also U.S. Public Building Services specifica-tions PBS-C.1 and PBS-C.2 have been drawn up to set norms for performance of these sound screens for speech privacy. In one privately conducted evaluation it was found that, whereas a sound reduction of only 4 to 7 dB was possible in an office with conven-tional low-form partitions, incorporation of 5 to 7 ft high fiber glass sound screens reduced intelligible frequencies of speech 8 to 14 dB.[5]

Industrial Noise Abatement

Early in the 1970's, the Occupational Safety and Health Act (OSHA) was passed by Congress. In addition to many other controls

Fig. 3-15. Open-office "sound screen" for partitioning and lowering speech intelligibility in conjunction with ceiling tiles, panels, or boards, and a masking sound system. Insert shows construction. 1. A metal septum to block sound transmission. 2. Fiberglas core (1 in.) on each side of septum to absorb sound. 3. Sturdy special Fiberglas sound diffuser (Glastrate) for abuse resistance. 4. Stain-resistant Dacron® Polyester fabrics. These fabrics are washable, color-fast, and fire-retardant (Class 25). 5. Extruded aluminum frame, fastened to septum for strength and stability. 6. Painted anodized aluminum kickplates for additional abuse resistance. 7. Top and side radii designed to minimize sound diffraction over edges. (*Courtesy Owens-Corning Fiberglas Corp.*)

set on industrial working conditions in this act, limits were set also on the lengths of time for which workers could be exposed to given noise levels.[6] These varied from 8 hr per day at 90 dbA to $\frac{1}{4}$ hr per day at 115 dbA. Because the human ear does not hear the same at all frequency levels, but is more sensitive in the 1000 to 7000 Hz range, a special "A" weighted scale was developed. This scale is used

to adjust automatically the readings in a sound-level meter emphasizing and favoring the frequency range 1000 to 5000 Hz where
human hearing damage is likely to occur. Hence the designation of
units in dbA indicates that sound-level readings have been made using
the compensating sound-level meter.[7]

After several years of activity on the part of industry attempting to
meet these OSHA acoustical requirements, there is a move being
made by that government agency to lower permissible noise levels on
a sliding scale beginning at 8 hr per day at 85 dbA instead of 90.
This is as yet an unsolved situation, with industry withholding complete commitment. The total cost to industry for this change is
estimated to reach billions of dollars.

One of the main difficulties in assessing the noise level to which a
worker is subjected is that frequently in a factory or workplace,
noise comes simultaneously from several emanating sources and at
different loudness levels. These must all be treated additively in calculating OSHA conformity. Effective ear-protecting devices for
workers are always available, and their use is required by any responsible industrial concern. Discomfort always accompanies their use,
however. None-the-less, industry prefers use of these individual
hearing protectors rather than insulation installation and physical
room changes, etc. Fiber glass manufacturers possess test equipment
for acoustical evaluations and measurements, and have a good record
of working with industry representatives and/or consultant acoustical
engineers in solving noise abatement problems.

In an industrial plant or operation, noise may be abated at its
source, along its path, or in the area of the receiver. Many fiber glass
products and product variations are available to assist in control and
reduction of industrial noise. These include roll-type wool products
and flat rigid or semi-rigid boards which can be used to line enclosures around noisy equipment, to erect screens or wall coverings, or
to fabricate baffles. Baffles may be wall-mounted or mounted and
hung as triangular or other geometric shapes. Pipe insulation reduces
or confines noise in many applications. Also, combinations of fiber
glass with other insulating materials form more effective noise
reducers. It seems that use of the same design or geometric configuration developed for the walls and ceilings of the anechoic
chamber for acoustical measurements would assist in abating industrial noise. Also fiber glass made with finer filament diameters as
used in aircraft and aerospace components (which see) is more effec-

tive than B-fiber in noise control and reduction, but also more expensive.

PIPE AND AIR-HANDLING INSULATIONS

Pipe Insulation

History and Evaluation

The very first mineral wool products produced (ca. 1840) were used to insulate steam boilers and piping. These fibrous products were acceptable because of their greater resistance to the high temperatures than organic material would provide. However, the need for binders, technical controls, and dimensional stability was probably very evident. In approximately 1900 in the U.S. (Alexandria, Indiana), rock wool as we know it today was first produced and used for thermal insulation in many domestic and industrial applications, including pipe covering.

Manufacture

Insulation of pipe forms is a natural adaptation, since the material can be fabricated into pipelike shapes with the I.D. of the insulation slightly larger than the O.D. of the pipe. The main method used is winding. Fiber glass pipe insulation is wound onto rodlike forms or mandrels directly from the forming units and prior to the time any curing action takes place in the binder. After wrapping, the cylindrical mold forms containing the uncured fiber glass are processed through a heated oven. Cure of the phenolic binder takes place in an ambient atmosphere of 425–450°F. For quality assurance, lengths produced are of the order of 3 ft or less. The cured segments are demolded longitudinally and finished to length specification by sawing. Other important parameters are glass fiber diameter and density, binder content (ignition loss), pipe diameter dimensions, and wall thickness. After slitting the pipe insulation longitudinally, facings are applied automatically (if required) and the products are packaged for shipment.

Forms Available

In the following are described the physical forms and general treatments available in fiber glass pipe insulations.

1. *Lengths.* The usual production sizes are 3 ft lengths but dimensions as short as 6 in. are supplied for use with smaller diameter copper piping.

2. *Diameters.* As stated, fiber glass pipe insulation is manufactured to fit the O.D. of the stated I.D. of standard pipe sizes. For iron pipe sizes, fiber glass pipe insulation is produced for the range $\frac{3}{8}$ to 36 in., and larger diameters may be produced for special purposes. For copper tubing or pipe sizes, fiber glass coverings are made for diameters ranging from $\frac{1}{2}$ to 12 in.

3. *Wall Thicknesses.* For iron pipe, the range of insulation wall thicknesses is $\frac{1}{2}$ to $3\frac{1}{2}$ in. For copper tubing, wall thicknesses range from $\frac{1}{2}$ to 3 in. Heavier wall thicknesses up to 8 in. or more may be applied by using multiple layers (nesting). Tables have been prepared to show nesting sizes, based on combined wall thicknesses and diameters.

4. *Jacketing.* The resiliency of phenolic resin-bonded fiber glass is brought to good use in piping. All pipe sections are slit longitudinally through one wall and partially through the opposite wall. This provides a clam-shell-type action when the insulation is placed around the pipe, making assembly much easier than if the insulation is in two half-cylinders, as in the case of some insulations competitive to fiber glass. For some special purposes, one type of pipe insulation is slit through one wall only, providing greater resiliency or spring action, and substantially speeding application over the substrate pipe.

Facings or jackets are factory-applied to almost all fiber glass pipe insulation. They are designed to provide protection against various elements in increasing degrees, and may be described briefly as follows:[8] (1) kraft paper treated for flame retardancy and adhered to aluminum foil with embedded fiber glass-scrim fabric (main purpose—fire retardancy), (2) canvas to provide a painting base, (3) fiber glass fabric for completely UL rated hot and cold systems, (4) metal (aluminum or stainless steel, .010 to 0.016 in. thick) for exposures to weather or extreme physical abuse, and to provide the best resistance to condensation. No real vapor barriers are required immediately adjacent to the piping to which the insulation is applied.

Almost all jackets are provided with some sealing means; either a lap, which may have a self-sealing adhesive, or with joint sealers and end seals.

Rapid fabrication at elbows is provided by a new technique based on a separately molded, folding PVC plastic contoured part that fits and closes around the elbow (see Fig. 3-16A). A blanket insulation napkin is first wrapped around the pipe.[9] This procedure eliminates costly, time-consuming hand cutting and fitting for insulating elbows. Some insulations, mostly unfaced, require other over-applications such as asphalt-impregnated fiber glass mats, banding for metal sheathing, and others.

5. *Temperature Resistance.* The usual upper temperature limit of 450 to 500°F has recently been extended to 650°F by virtue of minor but significant manufacturing innovations.[10] A minimum wall thickness of 2 in. is recommended for the 650°F applications. Ventilation must be provided during initial exposure above 500°F to remove fumes generated by partial outgassing of the binder ingredients. Temperature resistance to 850°F for pipe insulation

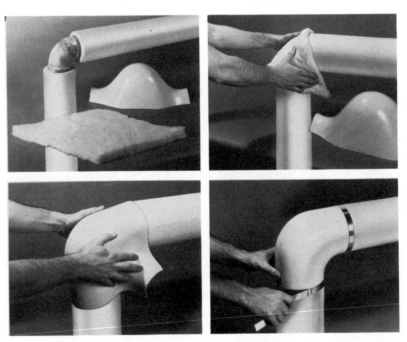

Fig. 3-16A. Rapid method for connecting straight fiber glass pipe sections at elbows using contoured foldable molded PVC shield. Blanket-type insulation is folded for encapsulating under the shield. This method eliminates costly manual cutting and fitting. (*Courtesy Zeston Inc., Subsidiary of Johns-Manville Sales Corp.*)

Fig. 3-16B. Placement of 850°F fiber glass insulation on iron pipe in a chemical plant. Since no facing is furnished, protective weatherproofing is post-applied. (*Courtesy Certainteed Products Corp., and Dow Chemical Co.*)

has also been accomplished using fiber glass together with a special resinous binder.[11]   See Fig. 3-16B. This material is supplied unjacketed, and the appropriate indoor or outdoor facing must be applied in the field.

6. *Fire Retardancy.* Almost all materials excepting some unfaced products comply with the requirements of UL Specification No. 181, Class 1 for fire hazard classification. The 850°F material, unfaced, does comply. Class 1 requires that the insulation material will not contribute to the spread of fire nor liberate excessive smoke (flame spread equals 25, and both fuel contributed and smoke developed equal 50 in this rating).

7. *Accessories.* A full line of accessory products and materials has been designed to enhance the function of fiber glass pipe insulation.   These include coatings, adhesives, outdoor and indoor vapor barriers and breather mastics, pipe supports and vibration isolators, and reinforced plastic fittings.

Properties and Performance

Uses. Fiber glass pipe coverings are generally used to insulate against both heat loss and heat gain (hot and cold applications, respectively). They may be used both indoors and outdoors and in many climates and ambient conditions, employing the proper facings as described. The types of substrate pipes insulated constitute iron (largest percentage of business), copper pipe or tubing, aluminum, and fiber glass reinforced plastic pipe if specified. Application may be either domestic, commercial, or industrial, and specific examples of applications are: *Heating*—high, medium, and low pressure steam lines, condensate returns, hot water dual temperature (4 lines), hot oil and reactors under 850°F; *Air or Water Conditioning* —chilled water, drains, brine lines, make-up water; *Plumbing*—cold water, hot water.

General Properties

1. *Weight.* Fiber glass pipe covering is light in weight, which makes it easy to handle and apply. The minimum density is 4.0 lb/cu ft nominal, and slightly higher for the higher temperature forms.

2. *Resiliency and Strength.* Pipe insulation exhibits excellent resistance to deformation. Properly jacketed material when installed will withstand full foot pressure of a man climbing or standing on it. Resistance to compression increases with percentage deflection but decreases with rising temperature. As an example, 6 lb/cu ft material at 450°F exhibits the following: at 5% deflection—1 psi resistance; 10% deflection—2.7 psi; 25% deflection—6.2 psi.[12]   The resilient property of fiber glass pipe insulations is also responsible for prevention of the edge-breaking or crumbling for which nonfibrous mass or bulk insulations have a tendency. Hence, the fiber glass insulation obviously allows for tighter joints, prevention of waste and avoidance of patching costs, thereby contributing to overall lower material costs.

3. *Resistances and Appearance.* Glass fibers per se are incombustible, will not burn, and will not soften below 1250°F; however, the binder material is an organic chemical. Glass fibers will not shrink, swell, rot, nor attract rodents or vermin. They do not attract, breed, or promote the growth of fungi or bacteria. There is no

acceleration of the corrosion rate of iron, copper, aluminum, or other pipe materials in contact with glass fiber insulation.

As regards appearances, the jacketed and/or metal-clad surface coverings present an orderly and clean appearance. Laminated kraft and fiber glass cloth facings may be painted. Fiber glass cloth facings are not recommended for damp, unventilated areas due to tendency for growth of mildew in the discontinuities. All jacketed pipe insulation may be easily maintained and wiped clean with a damp cloth. If proper thickness has been designed into the system, there is little or no tendency for condensation and concomitant drippage.

Specific Properties

1. *Thermal Conductivity*. The low thermal conductivity of fiber glass insulation (see Thermal Insulation—Homes) is due to the extremely fine fibers attenuated and their ability to trap and encapsulate a myriad of infinitely small air spaces or cells which act to retard the transfer of heat. Low thermal conductivity is further enhanced by the predominantly lateral orientation of fibers on the forming chain. This contributes also to greater mechanical strength in the circumferential direction as the insulation is wound onto the cylindrical forms. In Fig. 3-17 is shown thermal conductivity values over the partial "use" temperature range for fiber glass pipe insulations of three temperature limits: 450, 650, and 850°F. Plots for bulk hydrous calcium silicate and polyurethane foam pipe insulations are also included for comparison. The fiber glass exhibits a lower coefficient of heat transfer than that for calcium silicate in the temperature range up to 400°F, but shows greater rate of rise in conductivity above that temperature. Polyurethane foam possesses low K values, but is limited in temperature of application.

Related to the thermal conductivity values are the economical material thicknesses to be used under specific operating conditions. Economic thickness is defined as that thickness which generates minimum life-cycle cost for the owner. It is based on several business cash-flow factors. (see Fig. 3-11B). To save time in calculations, economic thicknesses have been tabulated for fiber glass pipe insulation covering heated pipes. Thicknesses required range from $\frac{1}{2}$ in. for 0.50 in. pipe size operating 100 to 200°F, to 5 in. thickness for 36 in. pipe size operating up to 500°F.[13]

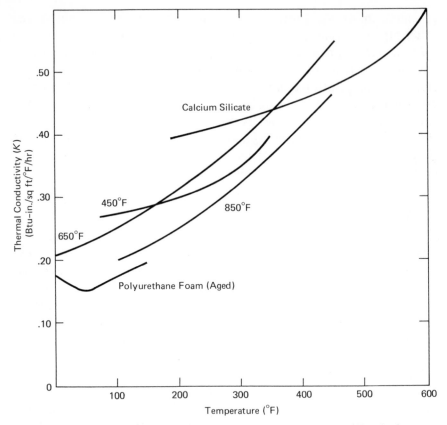

Fig. 3-17. Curves showing variation of thermal conductivity (K value) versus temperature for three fiber glass pipe insulations with end-use temperature limits of 450, 650, and 850°F. Curves for bulk calcium silicate and polyure-thane foam are included for comparison.

2. *Moisture Absorption.* 0.2% by volume after 96 hr @ 120°F, 96% RH.

3. *Specific Heat.* 0.20 Btu/lb/°F.

4. *Alkalinity.* pH of insulation = 9.

5. *Capillarity.* Negligible after 24 hr exposure.

6. *Expansion and Contraction.* Changes in linear and circumferential dimensions with temperature are well known for metals. For every 100°F increase or decrease in temperature metals normally used in piping systems would expand or contract respectively by the following approximate amounts in inches per 100 ft of length: iron pipe—0.66 in., copper—1.134 in., aluminum—1.484 in. Over

this same temperature interval of 100°, fiber glass insulation would expand or contract approximately 0.57 in. Hence it can be readily observed that not too many shear-plane or stress difficulties would result when fiber glass insulation was used with iron pipe. Iron pipe is involved in the major portion of all applications of pipe insulation.   Only differential temperatures and gradients cause problems.

As regards circumferential expansion or contraction the actual linear distances involved are negligible. In either case, the resiliency of the fiber glass material is instrumental in "giving with the blow," and rapidly relieving any stresses induced due to expansion or contraction.

Only in a very limited number of specific applications is it necessary to install expansion joints in the fiber glass insulation coating.

Specialty Applications—Combinations of Fiber Glass and Polyurethane Pipe Insulation.   These are beneficial due to the lower thermal conductivity of polyurethane, especially in the cryogenic region. Polyurethane is not servicable as an insulation medium above 150°F. It is effective at temperatures as low as -320°F, however. With the combined insulations, problems exist due to condensation when the pipe is at temperatures as low as -320°F and the ambient air outside the insulation is 90 or 100°F, 80 to 90% RH. Tables of recommended thicknesses to prevent condensation have been worked out and are available.[14]

Applicable Specifications
Fire and Smoke Hazard Classification
  Composite rating—insulation plus
  jacket plus adhesive   ASTM-E84
                                                      NFPA-225
                                                      UL-723
                                                      NFPA-90A

Jacket Properties (Nonmetallic)
  Water vapor permeability   ASTM-E96
  Beach puncture   ASTM-D781-63T
  Tensile strength   TAPPI-404-TS-66
  Mullen burst strength   ASTM-D774-67T
  PVC jacketing   FED SPEC L-P-535C
                                                      Comp. A Type II Grade G.U.

Government Specifications   MIL-I-22344B
                                             HH-I-558B Form D,
                                             Type III, Class 12 and 13
                                             MIL-I-24244
                                             HH.B. 100B, Type I and II
                                             HH.B. 1008

Insulation for Air-Handling Systems and Ducting

Introduction

The use of fiber glass insulation on sheet metal ducts and plenums was a natural consequence of its successful use as a thermal insulation on hot and cold piping. Use on ducting was given impetus by the rapid growth of improved heating and air conditioning systems. A further bonus was the contribution of the acoustical properties of fiber glass to noise elimination when used on both outside and inside ducting. Angular "crimping" and/or "breaking" of large sheet metal surfaces or panels is, of itself, not sufficient to dispel sounds generated by thermal changes and vibrations due to turbulent air flow within the ducts.

Further, when the more rigid board-type products became readily available, together with the science of affixing suitable protective surfacing materials, a natural technological consequence was the use of fiber glass alone to fabricate complete duct systems. In this segment are described in succession (1) use of fiber glass board products to insulate both the outside and (2) inside of sheet metal ducting, (3) use of faced roll goods for wrapping metal and similar air ducting, and ultimately, (4) the use of board products and the like in fabrication of air handling systems which are singularly fiber glass. Properties and performance data are presented for confirmation and understanding.

External Duct Insulation

Flexible blankets, semi-rigid and rigid board stock are all used for application to the outside surfaces of sheet metal plenums and ducting. Flexible materials have the advantages of greater facility of installation, bending readily around corners, etc. Board products must be fabricated, grooved, or notched to fit around corners. (See composite illustration, Fig. 3-18A & B.) Exterior facings supplied comprise the following: (1) none, (2) aluminum foil-scrim-kraft paper, (3) foil-scrim-plastic, and (4) glass cloth with a vapor barrier membrane, plus several variations of the above. Material densities vary between 1.0 and 6.0 lb/cu ft, and thicknesses up to 4 in. are fabricated, although job thicknesses to 10 in. may be required for the higher temperature forms and use above 500°F. Multiple layers require broken spacings.

As an example of thickness required, if insulation $3\frac{1}{2}$ in. thick is applied to a metal heating duct, for which the internal operating

Fig. 3-18A. Application of faced insulation to external portion of a sheet metal duct by impaling over welded-on studs and fastening with speed clips. (*Courtesy Johns-Manville Sales Corp.*)

Fig. 3-18B. Bending of sheet metal during duct assembly after the air-flow resistant fiber glass has been adhered to serve as an internal duct liner. (*Courtesy Johns-Manville Sales Corp.*)

temperature is 200°F, a cold-surface temperature of 86°F will be maintained on the insulation (80°F ambient, still air, full-time commercial operation).   Calculations of thicknesses required based on differential surface temperatures, densities, etc., are well detailed.[15]

Normal temperature resistance limits for fiber glass insulations are −60 to +450°F.   Some kraft type and other facings are limited to 250 or 350°F maximum exposure.   One type fiber glass board with a special proprietary binder incorporated provides service to 850°F (no facing, but metal or canvas applied over the cold side). The method of fastening comprises: (1) spot-welding pins or studs onto the metal surfaces projecting outward, (2) impaling the insulation, and (3) securing with washers or speed clips.   Recommended spacing for pins and studs is a maximum of 16 in. centers, starting at not more than 4 in. from any edge.   Specifications for thermal conductivity and fire hazard classification of these materials are similar to those given for fiber glass piping (see above).

Internal Duct Insulation

The success of fiber glass material used to line the outside of metal heating or air conditioning ducts prompted its use also as inner duct lining. An added requirement for duct liner applications is the ability to withstand the high velocities of the air being force-drafted and circulated within. Hence, most of the duct liner materials so used possess an added surface coating. This coating must also possess sufficient surface smoothness so that excessive air turbulence is not generated. These treatments are not applied to jacketing materials, but are incorporated into the insulation itself.

Understanding of the manner in which fiber glass and other materials affect an airstream requires attention to a new set of parameters—air friction correction factors for surface roughness. The technology for determining pipe size required for a given volume of air to be delivered at a specific pressure over a predetermined duct length is well documented in Chapter 21 of the ASHRAE Guide.

The manner in which the contemplated duct diameter must be changed if fiber glass is to be the design material is illustrated in Fig. 3-18C. Curves are shown for the several types of materials used for metal duct liners as well as for the all fiber glass ducting to be described later.[16] The effect of surface roughness (R) is to require enlargement of the contemplated duct size. The degree of enlargement is determined by multiplying the friction loss on the ASHRAE charts by the correction factor from Fig. 3-18C and readjusting the pipe size accordingly. Since round ducts are specified in the ASHRAE tables, requirements for rectangular ducts are directly related to cross-sectional area. Upper flow limits are indicated for each curve, although the materials are actually tested at 2.5 X specified velocity level.

The main difference compared to the exterior duct liners is that the interior type is limited to a maximum service temperature of 250°F. Duct liners are secured to sheet metal prior to fabrication with 100% fire-resistant adhesive. Then the metal may be fabricated with the insulation affixed (see Fig. 3-18B) If the internal duct facings are larger than 12 X 16 in., the adhesive for the fiber glass

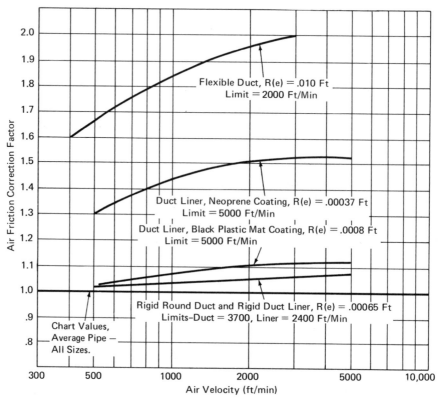

Fig. 3-18C. Curves showing variation of air friction correction with increase in air velocity for several types of fiber glass metal duct liners and for fabricated fiber glass ducts. Surface roughness R(e) values and maximum recommended air velocities are indicated for each curve.

must be supplemented by use of mechanical fasteners, and all leading edges heavily coated with the fire-resistant adhesive. For velocities in the range 4000 to 5000 ft/min, it is recommended that sheet metal fittings be fabricated and placed on the leading edges of the insulation to prevent delamination. Fire hazard classification, conductivity, and sound dispersement values are commensurate with those standards already given for similar fiber glass types.

Faced Insulation for Duct Wrapping

Several products in two major categories have been developed for wrapping around new or existing round, rectangular, or square ducts to provide insulation.   These products are provided with heavy-duty facings that prevent mechanical damage and also act as an effective vapor barrier, controlling condensation.

1. *Reoriented Insulation.* Flexible roll insulation in which the normal blown-fiber configuration has been systematically altered is manufactured in 3 lb/cu ft density and $1\frac{1}{2}$ and 2 in. thicknesses. The effect of the fiber reorientation is to provide a high compressive resistance of 200 psi at 25% deformation (room temperature). By comparison, 6 lb/cu ft standard board material deformed 25% offers compressive resistance of only 10 to 12 psi. Hence, the structural effectiveness of this material should be evident.[17]   (See Fig. 3-19.)

Two facings may be supplied: (1) a laminate of Tedlar® PVF film with asbestos felt and scrim plus metallized polyester film (vapor transmission = 0.53 perm.)*; and (2) the aluminum foil-scrim-anti-flame treated kraft barrier (vapor transmission = 0.02 perm.). Upper temperature limit of the material is 500°F. This material is utilized in applications requiring above-normal resistance to mechanical damage, and also where the extra-performance facings and high

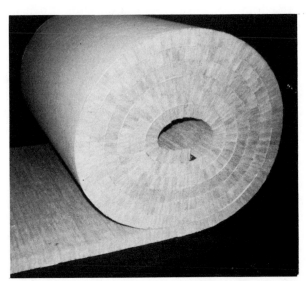

Fig. 3-19. Duct insulation in which the dominant direction of the fiber glass has been systematically reoriented to provide flexible roll-type insulation with exceptionally high compressive strength. (*Courtesy Zeston Inc., Subsidiary of Johns-Manville Sales Corp.*)

*perm. (permeance) = ratio of water vapor transmission to vapor pressure difference between the two facing surfaces; units = grams H_2O/ft^2–hr–in. Hg; see ASTM E96.

temperature resistance will compensate for the slightly higher economics.

2. *Faced Roll Stock, Normal Orientation.* Standard faced roll goods onto which special facings are applied also have been made available for duct wrapping. Densities are 1 to 2 lb/cu ft, and thicknesses are $1\frac{1}{2}$ and 2 in. Facings include (1) sheet vinyl (limited in use from 40 to 250°F) and (2) foil-scrim-treated kraft (to 350°F). Appropriate overlapping flanges are supplied when the facings are adhered to the fiber glass insulation, and these contribute substantially to ease of application to the duct. Again, use of welded pins and speed clips is recommended for larger expanses over 18 in.

Fabricated Fiber Glass Ducts

The fiber glass industry is slightly more than 40 years of age. The successes in insulation of metal ducting at only two-thirds that age led some astute researchers to ponder and investigate the efficacy of using hardboard fiber glass as the sole material of construction for an entire ducting system. The results were most successful due to the following: (1) easy handling and installation due to lightness of weight; (2) freedom from noise associated with expansion, contraction, and vibration of metal ducts; and (3) low maintenance costs and economy in energy savings. Other advantages will become apparent in the ensuing discussion of the various types of fabricated fiber glass ducting: (1) rectangular and other polygonal cross sections, (2) rigid round ducting, and (3) flexible round ducting. Rectangular duct is fabricated usually at the installation site, whereas rigid and flexible ductings are prefabricated and designed for easy assembly in the field.

Rectangular and Other Polygonal Ducts   Fiber glass board stock used to fabricate square, rectangular and like ducting, has only the outside surface faced with foil-scrim-kraft, while the internal side which contacts the air stream is left unfaced. The materials may be made into ducting for either air conditioning or heating systems, and the upper temperature limit when used in the latter is 250°F. In Table 3-9 are presented comparatively the salient properties of the three major types of board stock used. In addition, all three types are classed as FPM Roughness (e) = 0.00065 ft, correction factor varying between 1.02 and 1.08 (500 to 5000 ft/min velocity), ASHRAE guide.

TABLE 3-9. Comparative Properties of Fiber Glass Board Stock for Fabricating Duct Systems.

Type	Thickness (in.)	Stiffness or flexural rigidity EI[a]	Air velocity (ft/min) maximum	Internal pressure (in.-water gage) maximum	Limitations on duct dimensions (in.)	Thermal conductivity K, mean temp. = 75°F	NRC
Low pressure	5/8	Low	2000	0.75 in. pos. 0.50 in. neg.	20 in. max. 4 in. min supports every 3 in.	.23	.55
Standard duty	1	475	2400	2 in. pos. and neg.	None-larger than 120 × 120 in. if properly supported[b]	.23	.70
Heavy duty	1	800	2400	2 in. pos. and neg.	None if properly supported[b]	.23	.70

[a] Product of Young's Modulus (E) and moment of inertia (I), according to ASTM D1037-64.
[b] Positive pressure only.

Ducts may be fabricated by first grooving (vee or ship-lap) and then folding and sealing as illustrated in Fig. 3-20. Mechanical grooving equipment and installation are also shown. One manufacturer supplies premolded ship-lap ends for joining.[18] Being of higher density, there is less tendency in these molded ends for fracturing than cut or grooved ship-lap edges. Completion of the duct fabrication involves stapling and sealing over the flange provided on the originally adhered facing, sealing joints or branches (add-ons). Longer duct sections may be fabricated using the excellent mechanical closure channels available.[19] Supports must be provided at proper intervals, and many accessories are available such as turning vanes (fabricated within a duct elbow), dampers, etc. The technology has been well organized and documented, and is readily available for use. These systems are widely used in commercial and industrial building construction as well as in residential and mobile homes.

An excellent example of utilization of rectangular and associated (round and flexible, see below) duct is the all-air solar heat collector built recently in Denver, Colorado.[20] (See also Appliance and Equipment Insulation.)

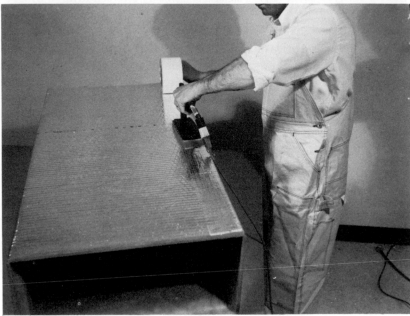

Fig. 3-20A. Grooving, folding, sealing, and applying protective sheet metal angle. (*Courtesy Johns-Manville Sales Corp.*)

Fig. 3-20A. *(Continued)*

Fig. 3-20B.   Mechanical grooving machine (*Courtesy Glass Master Co.*).

Fig. 3-20C. Installation of the finished rectangular duct in a dwelling under construction. (*Courtesy Certainteed Corp.*)

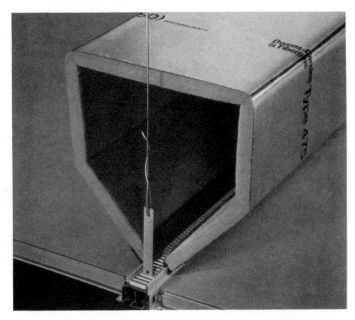

Fig. 3-21. Fabricated polygonal-shaped fiber glass duct installed above louvered channel. This unit is a portion of an integrated fiber glass ceiling system developed to provide coordinated air conditioning, heating, thermal, and acoustical insulation. *(Courtesy Owens-Corning Fiberglas Corp.)*

A 10-sided polygonal type duct may be fabricated to accommodate space limitations and produce an almost round duct.[21]   Also, a unique application of structured fiber glass ducting has been generated in the composite lighting-sound control-heating and/or air conditioning system for coordinated ceilings described earlier. (See Fig. 3-21.)

Rigid Round Duct Prefabricated rigid round duct is manufactured in a manner similar to that for pipe insulation. Thicknesses are usually 1 in., and the foil-skin-fire-control-treated kraft facing is applied to the outside. This pipe will withstand 3700 ft/min maximum air velocity at 2 in. positive or negative water gage pressure. Sizes produced range from 4 to 30 in. with 6 ft lengths for 4 to 18 in. diameters, and 4 ft lengths through 30 in. diameters. Ease of field fabrication is enhanced by the wide slip-joint at the ends. (See Fig. 3-22.)   Major uses are replacement of existing round metal piping, and also for new installations of heating and air conditioning

Fig. 3-22. Faced rigid round air-handling duct. Note the wide flange which together with improved sealing, round shape, and rigid fiber glass structure permits use at higher air velocities and internal static pressures. (*Courtesy Johns-Manville Sales Corp*.)

systems where space limitations favor use of round ducting. Better static flow in a round pipe system permits higher internal air flow velocities. Internal pressures to 8 in. water gage are possible using additional joint seals, and also employing the VAV system (variable air volume).

Round Flexible Ducting   A modern-day industrial-type "slinky" is used in fabrication of flexible round fiber glass air duct. This helical component makes possible a duct that can be installed through restricted areas and around structural obstructions, such as chimneys, piping, hangers, and others. This ducting eliminates turbulency generated by sharp bends. In fact, it can be routed into 180° direction changes without damage, and easily avoids any misalignment in connections. The wrapped-on fiber glass insulation is approximately $1\frac{1}{4}$ in. thick and is supported externally also by an aluminized plastic jacket or wrapper that provides a tough, seamless, fire-resistant vapor barrier. Limitations of 2000 ft/min velocity, 250°F temperature, and $1\frac{1}{2}$ in. water gage positive, $\frac{1}{2}$ in. negative are necessary. Diameters from 5 to 18 in. are supplied and lengths are 6 or 25 ft, except 16 and 18 in. diameters, only 15 ft maximum length. These elements may be shipped, compressed, and expanded at the site. The major commercial and industrial applications are use as branch ducts, but the material favors use in residential units also.

Specifications The following limitations are placed on the use of 100% fiber glass ductwork. They are not recommended in the following applications.

1. In or under concrete foundation slabs.
2. When air temperature exceeds 250°F.
3. Kitchen or fume exhausts or to convey solids or corrosive gases.
4. Above the recommended velocities or pressures.
5. Outdoors without proper reinforcement and weather protection.
6. Immediately adjacent to high temperature heating coils.
7. For vertical risers serving more than two stories.
8. In mechanical equipment rooms without adequate protection against damage.

APPLIANCE AND EQUIPMENT INSULATIONS

Introduction

Fiber glass contributes substantially to thermal and acoustical effectiveness and energy saving as an insulation material for both appliances and equipment. Probably the most logical delineation between the two marketing categories is size and manipulability. Two separate units may provide the function of refrigeration. One, either domestic or commercial, is an appliance if it is constructed so it can be moved. Another is categorized as equipment if it is built in place and may not be readily transferred or transported. In either case, several types of fiber glass insulations are available to satisfy the many and varied requirements.

Appliance Insulation

A list of appliances treatable with fiber glass insulations would include the following: domestic and commercial refrigerators and freezers, air conditioners, water heaters, vending machines, display cases, water softeners, insulated shipping containers, and many others.

Forms Available

Types of fiber glass developed to satisfy the requirements of insulation manufacturers include roll goods or batts. (1) Thicknesses vary from $\frac{1}{4}$ to 7 in. (2) Densities vary from 0.6 to 6.0 lb/cu ft. (3) Special treatments such as neoprene coatings or black coloration throughout may be incorporated for moisture repellency, resistance to degradation by airstream velocities, and resistance to dust accumulation. Manufacturers or specialty houses supply services of cutting parts to shape for facility in fitting and placement of installation by the appliance manufacturer. Treatments include cutouts, notches, corner cuts, slits and bevels, etc.

Product Properties

Generally, the overall properties which make fiber glass desirable for appliance insulation are: (1) good handleability and adaptability to expedite assembly; (2) varying degrees of compression resistance and flexibility to aid conformation to curved and irregular shapes; (3) sanitation—fiber glass felted material does not contribute to nor absorb and hold odors or toxic media; (4) stability—thermal and acoustical properties and physical dimensions do not deteriorate with time. Detailed properties are itemized and discussed as follows.

1. *Fire Hazard Ratings*. Requirements are more stringent for appliances than for pipe and ducts, with flame spread not greater than 20, fuel contributed limited to 15, and smoke development-maximum of 20 allowed (ASTM E84).

2. *Thermal Conductivity*. The 75° mean temperature K value is reduced slightly with increase in wool density, reaching a minimum of .22 at 2.4 lb/cu ft. Conductivities of 0.41 Btu-in./hr/sq. ft/°F are reached at a mean temperature of 150°F. This translates into the following thicknesses required: for a metal appliance substrate, over which insulation is applied, and operating temperatures of up to 450°F, 3 or $3\frac{1}{2}$ in. of insulation may be required, and the outside temperature may be 105°F.

3. *Sound Absorption*. Average noise reduction coefficients vary with thickness and change from 0.65 at 1 in. to 0.95 at 3 and 4 in.

thickness.   For appliances requiring sound reduction, the thicker the insulation can be made, the more favorable will be the results.

4. *Compressive Strength.* The resistance to compression at 50% deformation measures approximately 8 lb up to $1\frac{3}{4}$ in. thickness, and 10 lb for insulations to 4 in. thick. (OCF test method S-06 Ba.)

5. *Tensile Grab Strength.* Of importance in handing, this test shows 3.3 lb grab strength for $\frac{3}{4}$ in. material, 7.6 lb for 1 in; 12.9 lb for $1\frac{1}{2}$ in; and 18.2 lb for 2 in. (OCF Test Method S-010).

Installation of material in construction of appliance units is per-

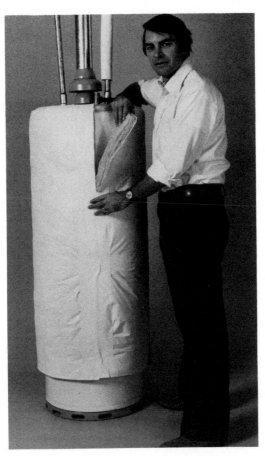

Fig. 3-23. Application of a prefabricated insulation kit to home water-heating appliance. Next to uninsulated attics, water heaters are purported to be the biggest source of heat loss and fuel waste. The kit accommodates all sizes of domestic water heaters to 80 gal. (*Courtesy Johns-Manville Sales Corp.*)

formed by laying in, draping and fitting over outer shells, and by impaling over welded pins. Another advantage of fiber glass is "fit," or the ability to be placed in confined areas without loss of thermal or acoustical properties. Also, the material, as previously demonstrated, has sufficient tensile strength to preserve its integrity and resist tearing when shells, etc., are placed over the insulation during assembly. An interesting adaptation to further augment performance and preserve energy loss is illustrated in Fig. 3-23. Although the appliance manufacturer designs as well and as economically as possible, more of the same helps. The extra insulation for home water heaters is distributed in kit form for do-it-yourselfers.

Miscellaneous

Other types of appliances, the operation of which is improved by incorporation of fiber glass insulations, include the following: dishwashers; clothes washers and dryers; disposers and other food-waste-treatment units; water softeners; air conditioners and heat pumps; oil, gas and electric furnaces; baseboard, room area or radiant-panel heaters; mixing boxes and air diffusers; home and other fire-places; beverage coolers and dispensers and other vending machines; ice makers and dispensers; storage bins; incinerators; incubators; lighting fixtures; laundry equipment; slow cookers; ironing board pads; insulated picnic bags, picnic coolers and food warmers; radio and television cabinets and stereo hi-fi speakers; juke boxes; and others.

Equipment Insulation

Types of commercial or industrial equipment whose function is enhanced by application of fiber glass insulations include the following: (1) heating equipment such as ovens, furnaces, boilers, solar heaters, hot-water generators; (2) cooling equipment such as large stationary air conditioners, walk-in coolers, reach-in coolers, through-put food-processing freezers plus other low temperature specialty equipment; (3) tanks and other storage facilities. The several types of fiber glass available to satisfy the service demands of equipment applications are described below, pointing out their most salient features and performance advantages.

Standard Roll-Type Insulation

Regular fiber glass material of the type used in building insulation has found ready use in equipment. The higher thicknesses and densities are employed primarily. Figure 3-24 shows adaptation to a solar heating unit. It is necessary to pretreat the insulation at 400°F or higher to out-gas the organic thermosetting binder. This pretreat-

Fig. 3-24A. Artist's rendering of a liquid-system solar energy collector showing panels angled and faced in the proper orientation.

Fig. 3-24B and C. The two inset views show placement of the insulation behind the absorber plate and also a schematic cross-sectional diagram showing the main components. The insulation is pretreated by heating to 400 or 450°F. This drives off a portion of the organic thermosetting binder, and prevents fogging of the glass panels and heat-collecting system due to the high in-use temperatures developed. Primary collection medium is a glycol-water mix. Heat is transferred through intermediate exchangers, and absorption-type air conditioning may also be designed in. (*Courtesy Libbey-Owens-Ford Co.*)

ment removes approximately 4% of the binder, preserves the integrity of the fiber glass, and helps retain its fire hazard classification. It also prevents internal fogging.

Dual-Density Insulation

Nonmembrane facings of either additional felted wool of higher density, or of blown mat types may be applied to the surface of standard insulations. Also, neoprene or other type of accessory after-treatment may be applied. Overall thicknesses from $\frac{1}{2}$ to 2 in. are supplied. Performance advantages are: (1) superior noise absorp-

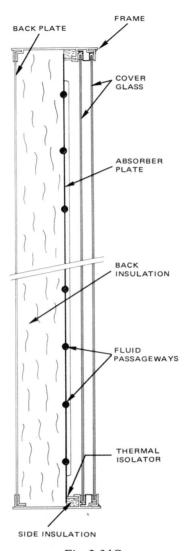

BACK PLATE

FRAME

COVER GLASS

ABSORBER PLATE

BACK INSULATION

FLUID PASSAGEWAYS

THERMAL ISOLATOR

SIDE INSULATION

Fig. 3-24C.

tion ($\frac{1}{2}$ in. = NRC 0.55, 2 in. = 0.95); (2) durability against mechanical and air erosion (possible-use velocity (ft/min) = 4000 to 4500); and (3) low thermal conductivity (mean 75°F K value = .48 for $\frac{1}{2}$ in., .13 for 2 in.); (4) fire hazard classification = 20-15-0. With a maximum operating temperature of 250°F, this material finds application in air conditioning and heating equipment.

Mechanically Bonded Mats

For special blanket types for insulating ship turbines, marine, industrial process and power generating equipment, plus furnaces and ovens operating to 1200°F, a product has been "borrowed" from the continuous-filament type fiber glass (see Chapters 4 and 5). Chopped continuous fibers are deposited randomly on a moving belt and "needled," mechanically bonding them into a highly integral, handleable product. Having no organic binder present, this material may be used up to temperatures approaching the thermal endpoint of the glass (1200°F maximum). It is nontoxic and nonodorous, also resisting damage due to vibration in the end-use application. This material is produced in thicknesses of $\frac{1}{2}$ and 1 in. and the corresponding product densities are 7.5 and 11.3 lb/cu ft, respectively.

Thermal Insulating Wool

A higher density, unfaced fiber glass insulation with thermosetting resinous binder has been developed expressly for thermal insulating up to 1000°F. Blankets (roll goods) and batts are produced, and show respectively, the following thicknesses required for insulation at 1000°F: blankets—8.5 in., batts—5 to 6 in. This is a product with low compressive strength, but in situ applications are highly successful (see Fig. 3-25).

Fig. 3-25. Use of unfaced thermal insulating wool to insulate a heated oil tank. A weatherproof coating is post-applied. (*Courtesy Certainteed Products Corp.*)

Mineral Fiber Board Insulation

A block-form insulation, made in finite sizes up to 36 × 24 in., and 1 to 4 in. thick, is composed of mineral fibers bonded with a clay-type binder. Density is 15 lb/cu ft, and the material possesses both high strength and resistance to fire sufficient to permit an end-use temperature of 1900°F. Thermal conductivity K values range from 0.36 at 200°F to 0.90 at 1200°F. The excellent high temperature performance makes this material advantageous for insulating boilers, tanks, and other heated vessels.

Double Mesh-Faced Insulation

To preserve integrity and make handling easier in installation, an insulation product is made with metal fabric securing both surfaces. The type fabric may be either galvanized 1 in. wire mesh, or copper-bearing metal lath $\frac{9}{16} \times \frac{7}{8}$ in. diamond-shaped openings, or combinations. This material is supplied in blanket form up to 24 × 96 in. sizes. Density is 3 lb/cu ft and thicknesses up to 6 in. With the advantages of the wire mesh affixed, the insulation has higher mechanical strength and may be better cut, shaped, and fitted, especially around obstructions and irregularly shaped protrusions. The fibers will not dislodge around the periphery, permitting holding of square edges in installations. The main advantage of the wire-mesh holding faces is to permit flexing to contours such as the tank exteriors, and holding to prescribed thicknesses without creating thin spots. A maximum use temperature of 850°F is designed in, but binder deterioration occurs, and hence a slow initial heatup (2 hr) is desirable.

Metal-Jacketed Equipment Insulation

To provide an exceptional high-integrity insulation product for direct application to heated tanks (450°F), a corrugated metal facing was designed.[22] The insulation thickness may be up to 3 in., and the aluminum facing has predrilled holes for application over studs welded to the tank surface. To allow for complete coverage, the aluminum sheet overlaps the insulation by 2 to 4 in. The main advantages are rapid installation and good finished appearance, and a near-permanent, weather-resisting outer tank surface.

Miscellaneous

Other types of equipment in which fiber glass insulation is profitably used from the economical and functional standpoints include the following: laboratory equipment; insulation of screw, pump, and heat exchanger lines in spinnerette assembly for synthetic fiber production; overhead and gravity-type fire doors, office and other wall partitions; curtain walls; cryogenic and cold-storage systems; shrouds and engine compressors; aluminum siding; computer equipment cabinets; insulated camping and other recreational or industrial tents; exhaust hoods for laboratory and industrial equipment; caskets (as padding); insulated live-animal dwellings and fish containers for permanent installations or mass shipment.

INSULATION FOR VARIOUS TRANSPORTATION MODES

Vehicles or carriers used in the three main areas of transportation, land, water, and air (automotive, marine, and aircraft-aerospace) all use fiber glass insulations in various forms. The purpose of the insulation is either to improve or make possible the actual functioning of the vehicle, or to thermally and acoustically enhance the comfort of the occupants being transported.

In this section, use of fiber glass insulations is described for the three modes of transportation in the order stated above.

Automotive Market

The development of the use of fiber glass in passenger cars and trucks is of sufficient interest that space be allotted here to trace it. In fact, the technology for portions of it is so unique that it has spilled over into other areas, as has become evident (see Acoustical and Pipe Insulations). Described herein are uses of fiber glass in automotive topliners, hoodliners, van engine housings, and miscellaneous applications such as bus heater-panel protectors, side-cowl and floor insulations.

Automotive Insulation–Topliners

Early-on in the annals of fiber glass development (ca. 1947–53, a small California company, Vibra-Damp Corporation, later acquired by

Glass Fibers, Inc. of Ohio, developed a method of making resilient pads by placing layers of uncured, phenolic-bonded B fiber wool in a press, closing to a thickness approximately double that for complete compression, and curing between heated platens. The result was a product that had almost 100% dimensional recovery over a long cycle of flexing and release. It was a perfect vibration damping material, that is, provided the load applied was not too weighty. Unfortunately the material failed by self-abrasion of fibers and fractures at binder junctures in heavy-duty use after repeated cycling. Another corollary development centered around one-per-shot recoil pads or shock buffers for small rockets. The latter was a government requirement, and probably millions of units are stockpiled somewhere, never to be used.

So the use of fiber glass as a vibration-damping medium joined the ranks of some other many-called-and-few-chosen product applications.* Fortunately, however, this initial skirmish into fabrication of B-staged wool did not become relegated to complete oblivion, but actually laid the groundwork for a whole new series of automotive products. In 1952, fiber glass dash panels were molded by clamping uncured wool between inner and outer sheet metal molds and curing them by hanging on an overhead conveyor passing through a heated oven. Covered with vinyl fabric before use, these parts showed the great versatility and adaptability of the fiber glass material by combining wide thickness variations and complex, three-dimensional contours in the same part. Although the dash panels had the advantage of gradual absorption of and increasing resistance to rapidly applied impact loads without suddenly bottoming out, as did foam plastics, the latter did take over, however, and the dash panel application was short-lived. Next came experimentation for usage in car door and ceiling panels, and the technology was developed for molding uncured fiber glass wool between contoured, heated platens in a press operation with mass production capabilities. Requests for these items came directly from the automotive engineering groups as an SOS to free car manufacturers from the laborious, many-component assemblies of door panels and ceiling installations.

In ceilings particularly, a five- or six-step operation was necessary (all right on the production line) using skilled labor and involving

*See, however, U.S. Patent No. 3,736,215.

the following: cutting fabric at the site, sewing in along ceiling cross supports, stretching out wrinkles, gluing above doors, adding bead and mold trim, and finally steaming to tighten the fabric and remove all residual wrinkles.

It is easy to see for anyone who has witnessed the operation of a modern-day automotive assembly line, that this older method was undesirable because of large space required near the line for equipment and materials, and length of time consumed in installation.

Using the molded fiber glass components, storage space required is minimal, and they are snapped in place by a well-trained but not a trade-skilled worker. Hence the car production line is very desirably speeded up. See Fig. 3-26.

Manufacture and Testing of Topliners In the manufacturing process, automotive topliners may be molded directly on line or the wool may be stored in the uncured state for later processing. Storage life is approximately 1 week at room temperature.

Only light construction is required for the press equipment, and shaped aluminum molds are used with no fine surface finish necessary;

Fig. 3-26A. Glass fabric faced contour-molded automotive topliner being installed during assembly of a recent model passenger car. (*Courtesy Johns-Manville Sales Corp.*)

Fig. 3-26B.   Finished car showing positioning of topliner and coordination with other interior components. (*Courtesy Johns-Manville Sales Corp.*)

aluminum is desirable for uniform and rapid heat transfer to promote even temperatures on mold surfaces.

The presses are operated with air cylinders, and proper guiding is necessary to avoid any sideways thrusting as the mold is closed. Press cycles are in the neighborhood of 1 to $1\frac{1}{2}$ min.

Two facings are used with topliners: decorative fiber glass fabric, or a decorative vinyl-impregnated fabric-based material, either continuous-film or perforated.

The fiber glass cloth has a special discontinuous treatment applied to one side to promote adhesion with the phenolic resin, and permit passage of sound waves for desired acoustical conditions within the car. For this reason, the cloth is preferred to a continuous film material. This treatment also helps to prevent bleed-through of the binder. Being more refractory, the cloth may be molded directly with the curing wool at mold temperatures of 600°F (maximum part temperature reached = 480°F). All thicknesses of glass cloth are employed, but the most desirable are those possessing a pattern or texture that hides any surface or density irregularities in the wool.

Vinyl materials are limited to a maximum temperature exposure of 240°F. Hence, when vinyls are to be applied, it is necessary first to mold the fiber glass topliner, then lower the press temperature and recycle each part, adhering on the vinyl material as a secondary operation. Both the fiber glass cloth and the vinyl-impregnated fabric are passed through tensioning rolls prior to combining with the glass wool to remove any wrinkles. Flashing is removed by hand cutting in a trim fixture, because shear edges are not employed in the molding system.

There are fairly stringent limits on total topliner weights. If the stock is too thin or light, molding is difficult, and parts have inferior strength. If stock is too heavy, the parts will not cure properly.

Hence uniformity of wool thickness and density from the machine is highly desirable, and these are checked on-line with regular frequency. Figure 3-27 shows a chart used in converting gram-weights per square foot to density in pounds per cubic foot for various material thicknesses. Wool with original weights of 1 in., 2 lb/cu ft density are used for topliners and after molding convert to $\frac{1}{2}$ in., 4 lb density at the center and $\frac{1}{4}$ in., 8 lb density at the edges. Other on-line tests are fiber diameter (B- and/or C-fiber used) and ignition loss.

Off-line tests conducted during manufacture are the following: physical dimensions and proper location of design or styling elements; vertical alignment of top and bottom mold halves; skewness or off-centering due to shifting or turning of mold halves, bleed-through of binder through the cloth; wrinkles or nonuniform density of the molded wool; fiber-fly (fibers detached and readhered onto cloth facing); cloth or vinyl adhesion; percent resin in the cured wool; proper tensioning of the cloth or vinyl skin; yellowing or any other color change in cloth or facings; and finally hand or "feel."

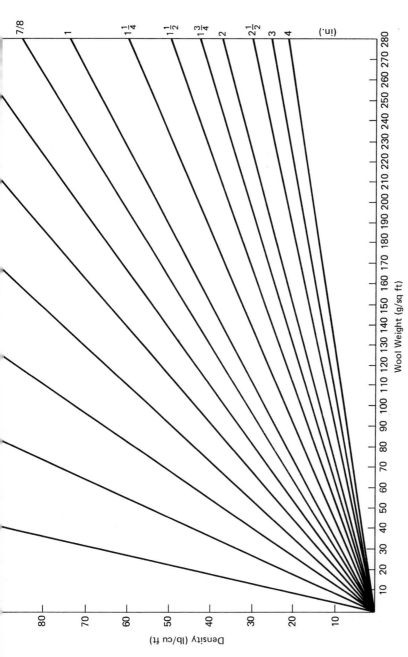

Fig. 3-27. Chart used to convert gram weights per square foot to density in pounds per cubic foot for various thicknesses of molded or regular fiber glass insulation.

Just as the automotive assembly lines must move rapidly for best economy, so must the topliner production lines proceed at the most feasible top speeds. Evaluation of the topliner product as both a technical and a mass production success can best be indicated by shop losses, which are almost negligible in manufacturing, and less than 1% rejected in the automotive plants.

Topliner Product Properties and Performance   The difficulties in automotive molded changeovers generated by the necessary 18-month lead-times pose no major problems for those concerned with furnishing topliners. The low tooling costs, process adaptability, and uncomplicated start-up have actually been beneficial to the automotive industry, especially in times of model changes. The versatility of the topliners favors coordination with all other related automotive design factors.

A topliner functions as both a thermal insulator and a sound deadener. The outside surface contacts the metal roof of the car and further prevents road and other external noises from entering the driving compartment.

Besides lending an aesthetic and pleasing look to the car interior, and helping keep the car clean, the topliners provide wide latitudes of styling and design freedom. They are structurally arched to snap in place, and there is no static fatigue or long-term cold-flow which alters their ability to remain in place. Styling interest lines may be easily incorporated and functional elements such as recessed areas for protective shoulder straps, sun or window visors, or relief for rollover bars may be readily designed in.

Topliner weight is very constant, and its lightness assists the car engineer to economize on weight. Metal car bodies grow or shrink due to thermal expansion as day-to-day environment or seasons change. The dimensional stability of the topliners plus their semi-rigidity and strength take up these changes without difficulty.

If a wider roof span is required by a larger car, slighty increased thickness may be designed for the topliner to avoid possiblity of fracturing or flexural failure. A slightly thicker cross section improves acoustical performance.

Acoustical testing is performed on a frequent spot-check basis by the receiving automotive manufacturers according to ASTM C384.

Fibers made using the flame attenuation process provide superior performance in topliners due to their greater length and better molding characteristics.

Although no flame retardant is used in the base binder resin, topliners have an excellently wide range of safety, as determined by Federal Specification DOT 302.

Automotive Insulation—Hoodliners

Hoodliners provide acoustical protection only, over the engine and under the engine hood, and are not used for thermal insulation. As are the topliners, they may be molded directly on the fiber glass wool production line. The thermoset binder used is colored black to make the parts blend with the inevitable discoloration from road dirt, etc.

Parts are molded for specific car designs, and no general-purpose units are produced. Wool thickness is slightly greater than that for topliner parts, approaching $\frac{3}{4}$ in. A wide flange around the periphery is compressed to $\frac{1}{8}$ in. to seal the edges and lessen the amount of dust generated in handling. The front or leading edge is treated with an elastomeric material to prevent airstream damage to the hoodliner while in service. The bottom or exposed side has no facing, however. Thin portions may be molded into the hoodliner panels to accommodate air-cleaner housings and other elements. Also, mounting holes are punched out in the post-molding trim operations.

Automotive companies check dimensions in test fixtures and frequently spot-test acoustical performance. Although the automotive hoodliners are technically fully functional, a large portion of their value is sales oriented. The fiber glass hoodliners provide real visual evidence to a prospective car owner that the manufacturer is taking positive steps to insure a quiet ride for the car passengers. See Fig. 3-28A.

Automotive Insulation—Molded Engine Housing Insulators for Vans

Uncured fiber glass wool may be fabricated by press-molding into complex three-dimensional shapes. The overall dimensions are approximately $2\frac{1}{2}$ ft × 2 ft × $1\frac{1}{2}$ ft. Facings may be co-molded on the outside or inside surfaces, although the housings are usually contained

Fig. 3-28A. Finished molded and trimmed automotive hoodliner ready for installation. (*Courtesy Johns-Manville Sales Corp.*)

behind an outer plastic shell or cabinet. These parts cover the portion of the engine that protrudes into the passenger compartment in modern-day vans.

The wool is compressed to $\frac{1}{2}$ or $\frac{3}{4}$ in. thickness in the body of the van housing, and again to approximately $\frac{1}{4}$ in. at the edge with no major flange, so as to eliminate minor dusting in handling. There is little drape to the uncured material in molding, so some wrinkles are induced. These are no major problem, however.

These elements serve four major functions: (1) thermal insulation, keeping excessive engine heat out and permitting uniform cab temperatures in both heating and air conditioning; (2) acoustical sound deadening, suppressing most of the engine noise; (3) cleanliness, because the inside added facing is a heavy, puncture-resistant aluminum foil directly molded on in the pressing operation; and (4) electronically insulating, because the heavy foil is grounded to the car chassis and bleeds off radio interference waves generated by the ignition system. In Fig. 3-28B is illustrated a van housing and plastic cover. In this model, there is no front facing, but the aluminum foil coating on the engine side is clearly evident.

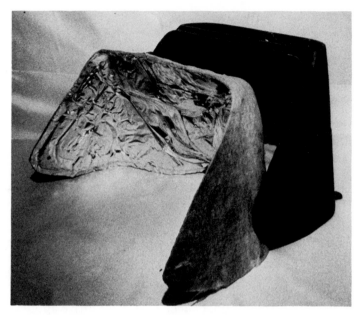

Fig. 3-28B. Molded fiber glass automotive van engine housing insulator showing adhered-on aluminum foil for specular heat reflection (engine side) and the external plastic cover (passenger compartment side). (*Courtesy Johns-Manville Sales Corp.*)

Automotive Insulation—Miscellaneous Components

Several additional molded components have satisfied critical thermal or acoustical demands in automotive construction. These are briefly described as follows:

1. Bus Panel Insulation. Large panels approximately 2 ft wide × 4 ft long are molded to help provide both thermal and acoustical insulation in the passenger compartments of commercial buses. Molded-in grooves are spaced to hold heater elements and foil may be molded in to increase insulation efficiency. These units are placed on the inside of the bus side panels.

2. Side Cowl Insulators. Performing strictly an acoustical function, the outside contour of these flat or structured components is molded to the precise cross section of the hollow space between front fender and inner fenderliner. The function is to block engine and road noise from entering the passenger compartment.

The fibrous material is black, and has a light scrim fabric molded in to protect and preserve panel integrity on the windward side.

Center thickness is maintained at approximately $\frac{1}{2}$ in., but the panels have outstanding strength and stiffness, resisting fracture and behaving as a self-supporting part. They are held in place in the cowl space by clip fasteners extending from the metal wall.

3. Floor Insulation. Molded fiber glass wool panels are placed under the car seats and floor sections to help seal in heat or conditioned air and also to block road noise. Foil may be molded in to further enhance thermal resistance. With the advent of the catalytic exhaust converter, higher temperature is developed under the car floor in areas close to the converter unit. High temperature fibrous insulations are discussed in a later section.

Many other molded insulating components for automotive use have been proposed and/or built experimentally and evaluated. These include ducting, dash-panel backup moldings, turbine intake liners, sun visor pads and many more. No doubt the versatility of these products will guarantee many additional successful automotive applications in the future.

Summary

The above characterization of current status of the use of insulations in automotive applications may be augmented by reiterating increased demand and market potential. The lightness of weight of fiber glass insulations makes them of extreme interest in weight saving, fulfilling obligations for energy saving. Also, the durability and resistance to slumping makes fiber glass insulations highly preferred over other types. Challenges are replacement of garnetted rag waste for under flooring, for the firewall, and for quarter-panels and various stuffers.

In addition to the automotive field, other varieties of land transportation have opened up as lucrative markets for fiber glass insulations. With revitalization of train travel and tourism, new passenger rail cars have been required, almost 100% insulated using fiber glass. It is interesting to note that insulation of railway cars provided one of the first major outlets for mineral wool heat-conserving materials in the period 1850–1900.

Marine Products

All vessels from outboards to aircraft carriers including submarines benefit from having some form of fiber glass built into their structure. The better outboards contain an insulation layer lining the motor shroud for noise reduction. The large ocean-going vessels employ fiber glass insulations as wall or room liners, and also for thermal and acoustical protection around turbines and similar gear. Products for these and for all the intermediate applications are described in the following.

Navy Hullboard

Dating from World War II this product was utilized as both room liner and finishing surface in construction of all types of vessels. Benefits were comparative light weight, extreme ease of installation, a tough, hard, durable, abuse-resistant surface, superior thermal performance, good resistance to fire, moisture, corrosion, and to slumping caused by vibration. Navy hullboard may be used plain as produced, or several types of facings may be incorporated. Unfaced material is usually used behind another type surface. Facings applied during manufacture or by specialty fabricating houses include heavy fiber glass fabric (most widely used), 3-mil aluminum foil, and others. Unfaced board, usually sanded smooth, has brattice* cloth or the equivalent applied as a covering following installation.

In use, the hullboard is applied over structural metal substrates of the ship (see Fig. 3-29). Its purpose is to provide an attractive, functional tactile surface for the interior of all living spaces, such as mess rooms, ward rooms, galleys, and other areas. It is also used as a protective thermal and acoustical insulating finishing material over bulkheads, beams, around decks, on shell plating, and ducts. The board with facings attached may be kerfed or slit to fit inside or outside curvatures and may be vee-grooved to fit snugly around the protruding structural elements. Fastening to the substrate is accomplished by using welded studs, impaling the board, and securing with speed clips, nuts and washers, etc.

The purpose of the glass cloth (most widely used facing) is to provide a strong, puncture-proof surface to protect the insulation from accidental damage. It is also an excellent base for paint in instances where the surface must be made totally impervious. The aluminum-faced material is used for special applications requiring ease of clean-

---

*Brattice cloth is a coarse, heavy, closely woven fabric made of jute, linen, cotton, etc., plain or twill weave, and treated for flame- and rot-proofing. It was originally used in mines for wind screens and control of ventilation.

Fig. 3-29. Installation of navy hullboard in construction of a ship. Finished hullboard is the panel at far left. (*Courtesy Johns-Manville Sales Corp.*)

ing, or radiation of heat. The unfaced, sanded material used for surfacing, of course, requires longer installation time if an extra cloth surface must be adhered on the job.

Densities of navy hullboard material range from 2.75 to 4.5 lb/cu ft, providing a product too stiff for roll-up. Hence, board sizes up to 4 sq ft are produced and sold. Thicknesses range from $\frac{3}{4}$ to 2 in. with 1 in. the most widely used dimension. Thermal conductivity at 75°F mean temperature is 0.23 Btu-in. A compressive resistance of 200 lb/sq ft is specified when the material is stressed to 40% of its original thickness. (The compression test is severe and conducted by: (1) determining the weight required to compress to 40% of original thickness, (2) immediately adding a weight equivalent to 2X the first weight, (3) releasing and then again determining the weight required to deflect to 40% of the original thickness.) Hence, higher resinous binder contents are usually required for these products to increase resiliency.

Considering fire retardancy, tested according to NFPA Specification 90-A, values of 25, 50 and 50 are permitted. Also, the hullboard material is subjected to passing the National Bureau of Standards heated-tube test, which is sometimes difficult due to the presence of the organic binder. Some regular fiber glass types have an "incombustible" rating, however.

Considering fire protection, each large ship, military or otherwise, is required to have a trained fire crew. This leads directly to a new concept in military ship building designated as SES (surface-effect ship). These high-performance vessels probably stem from the original, fast PT boats with a "planing" type hull, and are designed to be lighter and travel faster. New technology in providing insulation for these craft is built around use of higher temperature-resistant ceramic fiber. Temperature end limit for this material is 2000°F, or higher, and either none or very small amounts of organic binder are present. Hence, the naval architect is able to design a slightly larger ship which is still lighter in weight. Also, the higher temperature-resistant, ceramic fiber shows better performance in the heated-tube test, thereby not requiring presence of trained fire brigades aboard for any of the three daily watches (see also High-Temperature Fibers).

Other types of hull insulation are represented by unfaced or faced roll-forms for use behind other partitionings as both thermal and acoustical material. Thicknesses range from $\frac{1}{2}$ through 3 in. Facings are usually aluminum foil for a vapor barrier addressed to the warm side.

The "bonus" type for best acoustical insulation is a finer-fibered insulation (A- or AA-fiber) with NRC values of 0.45 for $\frac{1}{2}$ in. thick, 0.6 lb/cu ft density, to 0.80 for $1\frac{1}{2}$ in. thick, 3.0 lb/cu ft density.

Applicable specifications for hullboard are:

Rigid Board:
  U.S. Coast Guard No. 164.009
  Military Specification MIL-I-742c
Semi-Rigid Board:
  Maritime Administration Spec. 32 MA-3
  Military Specification MIL-I-22023
Blankets:
  Same as above specifications.

Acoustical Marine Fiber Glass Insulation

A 1 in. thickness of fiber glass insulation may be adhered to the upper surface of a hard, strong, perforated $\frac{3}{16}$ in. thick veneer-type material to form acoustical ceiling panels for use in staterooms, public places, crews' quarters, corridors, on bridge, etc. The composite performs as a decorative, sound absorbing, fire-resistant ceiling for all ships, since both the fiber glass and veneer are incombustible.

Applicable specifications are:

U.S. Coast Guard No. 164.009

Maritime Administration 32-MA-5

Marine Equipment Insulation

Excepting those products used for the SES programs, marine insulations described heretofore are designed for a temperature range from 40 to 250°F. Turbines and like marine equipment call for higher thermal endpoints and hence the following represent the group of products designed and produced for this end use:

1. *Felted Mineral Fiber Blankets.* These consist of mineral fiber felted between either asbestos fabric or stainless steel wire mesh. With stainless steel used, a temperature endpoint of 1100°F may be achieved. The mineral fiber may possess a density of 8 lb/cu ft and thermal conductivity K values at mean temperatures of 0.24 at 100°F, 0.40 at 400°F, and 0.63 at 700°F. Blanket forms are usually prefabricated for application to a particular piece of equipment (see Fig. 3-30). Thicknesses up to $4\frac{1}{2}$ in. are required to contain temperatures of 1100°F. Glass fiber may also be used as the encapsulated insulating material.

2. *Unbonded Mats or Batting.* Unbonded blown fiber glass wool felts or the "needled" chopped continuous glass fiber mats previously described are also used to insulate marine equipment.[22] The K factors for both materials are very similar: approximately 0.43 at 500°F. Thicknesses supplied are to 2 in. for the felted material and to 1 in. for the needled mat. Temperature end limits are: felted—1000°F, needled—1200°F. Both are excellent materials to use because of their drapability. Governing military specifications are MIC-I-24244 (BU Ships) Type 12E, MIL-I-164110, Type 2, and U.S. Coast Guard No. 164.009.

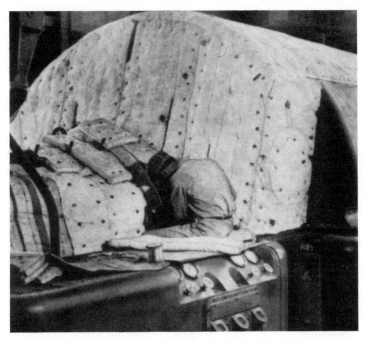

Fig. 3-30. Photo shows worker covering the surface of a marine turbine with felted mineral fiber insulation which has been prefabricated into specified shapes and encapsulated in asbestos fabric. High temperature limit is 850°F, and 1100°F if stainless steel wire mesh covering is used. (*Courtesy Johns-Manville Sales Corp.*)

Flotation Wool

This product is included here in the sense that a shipwrecked sailor floating on the water and being supported by life jackets containing a fiber glass flotation media is deservedly being benefited by a viable form of transportation. Fiber glass treated with a combined phenolic- and silicone-containing binder and cured at a temperature of 600–700°F (well above normal for standard-type insulations) possesses the capacity of repelling water and providing excellent buoyancy. This end-use performance is further enhanced by use of A-diameter fibers, trapping more air spaces, and providing greatly increased surface areas for repellency of the water. The high contact angle of the silicone material accounts for the water repellency, and hence, buoyancy.

Naturally, rigorous testing procedures were adapted for this material, since life saving and loss of life represent a short span of product performance having either high quality assurance or failure at its extremities. Although weighed quantities of the silicone-treated fibrous material were encased in rugged vinyl pouches to form a completed life jacket, testing was performed by placing the bare fiber in a pervious canvas bag so that it was in direct contact with water, and heavily weighted properly to evaluate its buoyancy. Test duration was 24 hr. The purpose was to guarantee 24 hr. safety in the water even in the event that the protecting vinyl covers were punctured by accident.

Although the contracts to produce this type material were fulfilled several years ago, the "flotation" technology has carried over into design and production of flight insulations. (See Aircraft and Aerospace Insulations.)

Aircraft and Aerospace Insulations

Introduction

A standard economical and technical criterion in the aircraft industry, especially in construction of large passenger and cargo jetliners, is that elimination of 1 lb of weight in design saves $1.00 in flight costs for every time the plane is airborne. This is not difficult to understand, and accounts for the very successful uses of fiber glass as effective, lightweight acoustical and thermal insulations in aircraft construction and operation.

These uses encompass two major areas: (1) incorporation of batts, blankets, and molded parts into the skeletal aircraft frame etc., as the main insulation medium; and (2) application of a specially developed, high temperature-resistant, thermal-shockproof material as the exterior skin of a newly developed space shuttle and orbiter vehicle. These two spheres of application will be discussed individually.

Aircraft Frame Insulation

Three forms of materials are used: (1) felts of AA-and B-fiber blown fiber insulation with plain or silicone binder for water repellency (see Marine Products); (2) molded AA-or B-insulation (supplied in uncured form for processing into specific shapes); (3) webs, mats, and thicker felts of Micro-Fibers® or high-silica (quartz) fibers of exceedingly fine filament diameter. The microfibers and fine-fibered quartz materials do not contain binder, nor are they supplied with any facings.

**TABLE 3-10. Comparative Thermal and Acoustical Properties of
Fiber Glass Used for Insulation of Aircraft.**

Material	Density (lb/cu ft)	THERMAL CONDUCTIVITY (K) AT TEMPERATURE OF: (Btu-in./sq ft-hr-°F)			Acoustical NRC value
		100°F	500°F	1000°F	
Micro-Fibers®	4.0	.23	.43	–	–
	5.0	.23	.42	.71	–
Micro-Quartz®	3.0	–	.50	.91	–
	6.0	–	.39	.68	–
AA-Fiber	0.6	.25	.77	–	.48 (½ in.)
	1.0	.24	.70	–	.72 (1 in.)
B-Fiber	0.6	0.30	0.99	–	–
	1.0	0.27	0.78	–	.70 (in.)

The AA-and B-fiber felts and blankets, usually containing the water-repellent binder, may be supplied plain, quilted, or between PVC, reinforced mylar, coated nylon, or glass fabrics.

The molded materials are fabricated by an end user, or some firm close to the end operation, and are supplied uncoated or unfaced. However, they may have some protective material applied at the time of installation for purposes of long-term protection against dirt, grease, moisture, etc. Comparative properties of the base materials are shown in Table 3-10.

As regards usage, the Micro-Fibers® and high-quartz felts are employed as nonconducting, separating media between reflective metallic shields in cryogenic environments, as thermal insulation fabrications for aircraft and jet engines, and in aircraft missiles and spacecraft as reinforcement for high temperature plastics applied in jet exhaust systems, nose cones, and aerodynamically heated surfaces. The temperature endpoint limitations are 900 or 1200°F for microfibers (depending on which base borosilicate glass has been used), and 1800°F for the high-quartz fibers. All components fabricated using these materials provide the desired low-light space-weight relationship.

The quilted AA-and B-fiber materials are fitted into the aircraft skeletal structure during construction. Economically, B-fiber is less expensive because of higher possible production rates, but AA-fibers provide superior thermal and acoustical insulation over B, especially at elevated temperatures. Large forms of B-or AA-fiber (2 ft × 6 or

Fig. 3-31A. Quilting loom into which AA-fiber glass blanket material is being fed together with facing material for both sides. (*Courtesy Flight Insulation, Inc.*)

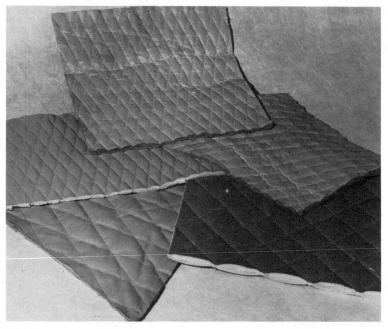

Fig. 3-31B. An assortment of various quilted insulation blankets. (*Courtesy Flight Insulation, Inc.*)

Fig. 3-31C. Installation of molded AA-fiber glass segments between ribbing just inside the outer walls of a large, wide-bodied jet aircraft. (*Courtesy Johns-Manville Sales Corp.*)

8 ft) are molded from uncured wool for direct installation into sidewalls between structural ribbing. Smaller molded forms are used in engine nacelles and other areas for thermal protection and noise reduction. Figure 3-31, a composite, illustrates (A) a quilting loom into which AA-glass fiber felt and facing materials are being fed, (B) an assortment of quilted products,[23] and (C) installation of molded AA-fiber panels between ribbing in a large jetliner, with the protective facing applied over.

Reusable Surface Insulation for Orbiting Space Vehicles

RSI Concept One of the many goals of the entire NASA space program has been the elimination of all waste. Firing a rocket to lift a capsule or other payload into space is akin to driving a multimillion dollar truck over a cliff immediately after it has delivered its first (and only) payload. Ablative systems do not completely or satisfactorily fulfill aerospace reuse requirements because they are too costly in turnaround time, i.e., replacement of ablated parts between flights is both costly and time consuming, particularly if the replacement material must be reapplied over large areas.

Hence, in approximately 1968 it was conceived to develop an orbiting spacecraft that could shuttle back and forth from earth to the cosmic environment, clad only with materials that would withstand the tremendous reentry heat load, and be able to become airborne again after an absolute minimum of ground time for refurbishing and surface parts replacement. Lightness of weight was of course paramount, and the goal was set at 100 shuttle trips into space and back.

The metal aluminum was selected as the substrate or primary structure to be protected to a maximum of 350°F by skins or sheathing materials designed to resist temperatures generated during reentry. These temperatures range from 600 to almost 3000°F. Tests and calculations revealed that over 70% of the vehicle surface would be exposed to temperatures up to 2300°F, and for this major area the reusable surface insulation was sought.

First materials considered were crystalline mullite (judged to have too few possible reuses and to be too great in density), and an overlapping, sheathed or plated structure of columbium metal (determined to require too long for complete development). Further investigation led to the use of a vitreous fiber material.

Since 1951 one of the prime contractors had been working with a fibrous all-silica RSI.* This material was selected because of its low thermal expansion, combining in the fibrous state with low thermal conductivity to provide best thermal shock resistance of any applicable ceramic. The properties of light weight with controllable varying bulk density were synergistically associated. Also, the material could be formed into components with aerodynamically smooth surfaces which also presented high resistance to moisture absorption.

The space shuttle design settled upon is illustrated in Fig. 3-32.[24] Inserts indicate isotherms in °F for the various surface areas of the spacecraft (temperatures generated during reentry). Size of the orbiter in relation to known commercial aircraft is of interest and is presented in Table 3-11.

The unit is powered for flight with three liquid hydrogen-liquid oxygen rocket engines clustered at the empennage, each with a 470,000 lb thrust. For vertical launching, the total thrust is increased to 2.65 million pounds by an added solid booster. A gener-

*Reusable surface insulation.

Fig. 3-32A.   An artist's rendering of the NASA-Lockheed space shuttle in flight, reentering the earth's atmosphere.   (*Courtesy NASA, Lockheed Missiles and Space Co., and Johns-Manville Sales Corp.*)

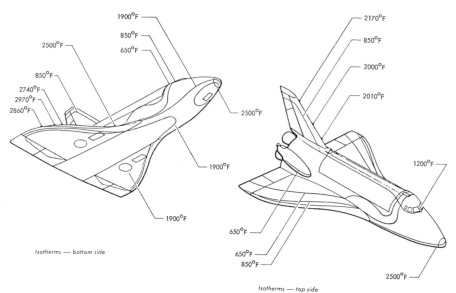

Fig. 3-32B.   Isothermal lines structured on the top and bottom surfaces of the space shuttle indicating temperatures reached during reentry. (*Courtesy NASA, Lockheed Missiles and Space Co., and Johns-Manville Sales Corp.*)

TABLE 3-11.

Type aircraft	Wing span (ft)	Overall length (ft)	Total weight empty (lb)
Space shuttle orbiter	76	122	143, 300
DC-3	95	65	16, 400
747-100	196	231	354, 060
Concord	84	202	174, 750

ous payload size is provided by a cargo bay slightly larger than that for a two-engine DC-9 jetliner.

Although approximately $5.0 to $6.0 billion were required for final development and construction in the years 1976 and 1977, each orbiter space-shuttle flight is estimated to cost only $10.5 million versus $30 million for a flight involving an expendable rocket.

Fibrous Surface Insulation Material, its Preparation, Application, and Performance   The base fiber for the RSI material is a blown Micro-Fiber,® filament diameter 1 to 1.5μ, made using a standard insulation-type glass (JM 475) composition.   By a special acid-treating process, the alkali is removed and the silica (SiO_2) content is increased from 58 to 97.5%. (See Filtration.)   This is the substance supplied for final fabrication of the space-shuttle surface elements that make up the protection system for the temperature range 650 to 2300°F.

Processing steps involved include the following: (1) the bulk fiber mass is fired to 2200°F to shrink the fibrous silica glass and eliminate microscopic voids; (2) a water slurry is made to establish good fiber distribution, and the mass is then dewatered by drainage, centrifugal casting, exerting a vacuum and/or pressure pumping; (3) with the water content at a minimum, colloidal silica (also 99.7% purity) is added, permeating the entire structure, and then the primary tiles (sized 15 X 15 X 6 in.) are cast; (4) these are humidity-stabilized at room temperature, dried in a microwave oven, and then fired at 2350°F for 2 hr to sinter the silica and provide a mechanical bond; (5) blocks are cut 6 in. square X 0.20 to 4.5 in. thick (use-thickness determined by exposure temperature and position on the spacecraft); (6) the density of the finished tiles is 9 lb/cu ft but may be increased to 15 or 30 lb/cu ft for areas subjected to possible impact damage, adjacent to doors, etc; (7) two types of boron-silicide glaze are applied, differentiated by end-temperature requirements of 600 to 1200°F, and 1200 to 2300°F, and the glaze is fired at 2260°F for $1\frac{1}{2}$ hr; (8) the tiles are abrasive-machined to within .015 in. of final dimensions or contour, and a final protective polymeric silicone coating is applied; (9) the tiles are prebonded into a block array with 0.050 in. gap between each, and a base filler strip between all tiles 2 in. and greater in thickness; (10) the tiles are adhered to the substrate using two layers of RTV silicone rubber separated by a thin padding of an elastomeric material to serve as a stress-relief layer.

Approximately 34,000 tiles 6 X 6 in. square are required to cover the 70% of the spaceship surface exposed up to 2300°F, and these provide an excellent shield with the following specific properties: (1) waterproof—although the glaze applied is not 100% crack-free, the cracks are not visible to the unaided eye, they do not propagate nor cause failures in the tile, and the applied polymeric coating assists in preventing moisture pickup and concomitant gain in weight; (2) SiO_2 has a low thermal expansion coefficient of 5×10^{-7} in./in./°C, and the tile will withstand repeated cycling from 2300°F to cold water without damage due to cracking or spalling, thus indicating that the space shuttle orbiter is good for 100 flights before replacement (reentry heat-pulse time persists for 10 min); (3) an extreme acoustical shock of 165 dB is generated at launching, and produces no difficulties in the coating; (4) as stated, tile densities up to 30 lb/cu ft may be produced for application to potentially damagable areas; (5) tiles can be repaired if scratched or damaged by application of pulverized glazing material through a heat or plasma-arc gun.

Remaining components employed to make up the complete thermal protection system are: (1) for nose cone and leading wing and tail edges, a laminated carbon-carbon composite* to provide heat resistance up to 2970°F (up to 300 sq ft of area); (2) a lightweight, elastomeric insulation for application to areas reaching 600°F or less.

Usage Programmed for Space Shuttle Orbiter   By the time this halting prose appears between hard covers, the orbiter will already have been subjected to eight landing tests after having been launched unmanned and piggyback from a NASA/Boeing 747. Soon after, if the decision-tree continues to branch in the "yes" and "go" directions,

*Manufactured by LTV Aerospace Corp., Dallas, Tex.

manned orbital flights will take place. Then, if successful, space transportation flights manned, and using the payload area, will follow.

Anticipated usage for these orbiter vehicles in space is first to remain aloft for periods from 3–4 days to 1 month during which any of the following could be accomplished: (1) monitor weather; (2) emplace new orbiting satellites and retrieve old or outmoded units; (3) mapmaking; (4) observe and study the earth for new mineral, agricultural, fishing, or energy fields using thermal receptors, etc; (5) experiment with manufacturing and scientific techniques using the cosmic vacuum environment of space; (6) resupply orbiting space stations; (7) efficacy as an instrument for military missions is obvious; (8) transport material and equipment to planned power stations in space which will transmit solar energy back to earth via microwaves.

HIGH TEMPERATURE INSULATION: REFRACTORY FIBERS

Introduction

The high temperature fibers discussed so far have been those made with the regular commercial insulating glasses or mineral fiber compositions, either with special binders or with the standard binder removed. Their maximum thermal endpoints are approximately 1200°F for glass, or up to 1800°F for the mineral fiber compositions. The fiberizable glasslike compositions of importance here are more refractory by comparison, and will withstand long-term usage temperatures up to approximately 3000°F.

Refractory fibers are produced from several different compositions of alumina-silica glassy melts and also from molten alumina, zirconia, and other metallic oxides. Also, as indicated in the Aircraft–Aerospace and Filtration sections, almost 100% pure silica fibers are produced by acid-leaching the alkali out of a glass after lower temperature fiberization.

There are three major factors that govern fiberization and usability of any glass melt, i.e., melting point, melt viscosity, and crystallization (tendency for devitrification). These are illustrated in Fig. 3-33. Part A shows the phase equilibrium diagram (temperature vs composition) for the binary alumina-silica system which forms the basis for most of the refractory fiber compositions employed in industry. Part B illustrates the relationship of fiberizing potential to melt viscosity as controlled by temperature.

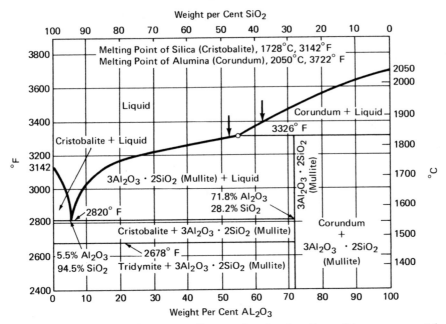

Fig. 3-33A. Alumina-silica phase diagram showing location of two commercial ceramic refractory fiber compositions.

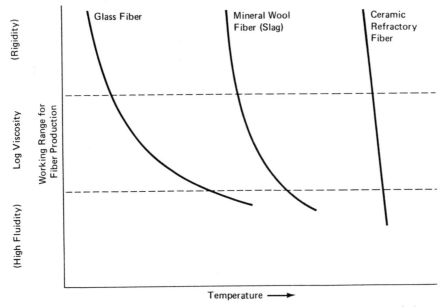

Fig. 3-33B. Schematic representation of the working viscosity range of glasses, mineral fibers, and refractory fibers illustrating the short fluid-to-solid cycle for the latter.

It will be noted that pure silica melts at $3142°F$ and pure alumina at $3722°F$. (Two other sources of refractory fiber, Cr_2O_3 and ZrO_2, not shown in Fig. 3-33, melt at 3600 and $4500°F$, respectively).

The two most widely used refractory fiber compositions are indicated by arrow in Part A. These are 52% alumina, 48% silica; and 62% alumina, 38% silica. The theoretical melting points of these compositions are approximately 3300 and $3400°F$, respectively, but due to melt viscosity and other considerations, temperatures in excess of $3500°F$ must be achieved to permit fiberization. Also, although $2820°F$ has been determined as their prime solidus temperature, practical operational temperatures (maximum) for these two refractory fiber compositions have been determined to be 2300 and $2600°F$, respectively. Actually, the percentage of crystallization is a major consideration, with only small amounts permissible.

The schematic illustration of viscosity and glass working or fiberization range versus temperature in Part B shows the long, slow rate of change of viscosity for an insulation-type glass versus the very narrow temperature difference between fluidity and rigidity for a refractory fiber composition.

This characteristic of refractory fiber compositions accounts for the large percentage of "shot" or unfiberized beads that are indigenous to their manufacture. The leaching of a lower-melting glass composition to remove the alkali fraction and leave a silica-rich fiber represents a successful industrial attempt to create a high temperature-resistant fiber without having to cope with the excessively, high and difficult-to-obtain melting temperatures, and their concomitant problems of shot, etc. (see Aircraft and Aerospace).

Hence, to summarize, practically all refractory fiber materials are fabricated by rapid air or steam, or mechanical attenuation of a stream or streams flowing from a glassy melt (mineral fiber process). The major production difficulties, in addition to those already described, are the commercial availability of refractories, fuel and firing systems, and associated accessory gear to turn out products of consistently acceptable quality and performance, and at a profitable economic level. The following will attest that this has been accomplished successfully by several firms.

The uniqueness and versatility of high temperature fibers may best be shown by describing the many forms available together with the applications and performance characteristics of each. Included are bulk fibers, blankets, boards, textile products, paper types, cast shapes, and other products which are the result of mixtures.

Bulk Fibers

Bulk ceramic fiber finds adaptations in the grades shown in Table 3-12. Generally, the fibers are white in color, possess good stability at the specified end-use temperature, have low heat capacity and hence do not "store" heat, have good thermal shock resistance (low thermal expansion coefficient), and are resistant to corrosive chemicals except for hydrofluoric and phosphoric acids and concentrated alkalis. They resist oxidation and do not deteriorate after becoming wet and being dried out. Thermal conductivity for several separately packed densities of bulk ceramic fiber to 2000 and 2300°F is shown in Fig. 3-34.[25]

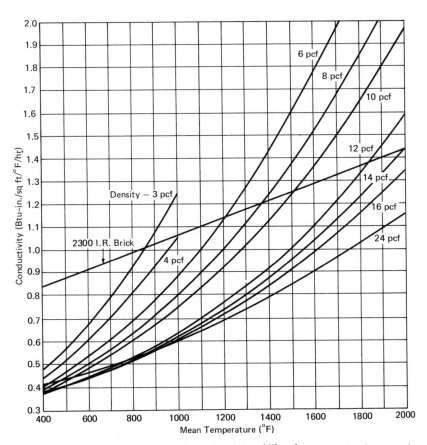

Fig. 3-34. Curves showing thermal conductivity (*K*) values vs mean temperature for different types and densities of bulk ceramic refractory fiber. A curve for insulating refractory brick designed for use at 2300°F is included for comparison. (*Courtesy Johns-Manville Sales Corp.*)

TABLE 3-12.   Properties and Applications of Bulk Ceramic Fibers.

Type fiber	Mean fiber diameter (μ)	Fiber length (in.)	Specific gravity (g/m^3)	Specific applications
Long fiber (Approx Al_2O_3—52%, SiO_2—48%)	Fine—8 Coarse—18	Up to 10	2.62	High temperature (to 2300°F) filtration and acoustical insulation; fluidized bed diffusing medium; raw material for high temperature textiles.
Bulk fiber	2–3	Up to 4	2.53	Resilient fill or packing material in high temperature expansion joints, kiln and furnace burner openings, and general high temperature packing insulation.
Washed fiber (most of unfiberized shot removed)	Fine—1.6 Coarse—2.3	To ½ To ½	2.53	Insulation for aerospace vehicles and used in ablative compositions; high temperature reinforcement for plastics; fine grade provides smoother texture for cast parts.
Milled fiber and chopped fiber	2–3	Chopped—300 μ Milled—14 μ	2.53	Additives for reinforcement, as compact filler insulations.
Higher alumina bulk fiber (Al_2O_3—62%, SiO_2—38%)	2–4	To 1	2.6	Can be used continuously at 2600°F; duplicates uses described above.

Felts, Blankets, Boards

Refractory fibers are fabricated into a series of flexible felts and blankets and into semi-rigid and rigid boards. Small amounts of organic and inorganic binders are used for product integrity. Other type bindings involve "needling"* to gain mat-product strength while eliminating added chemical binders. The inherent high temperature acoustical and thermal properties are of great help in industrial problem solving.

Felts and mats range from $\frac{1}{4}$ to 2 in. thickness, and up to 8 lb/cu ft density. Major uses are furnace insulation, firewall protection, packing for stress-relieving of welds, skid-rail insulation for heat-treating ovens and kilns, coverings for thermal insulation while transporting hot ingots, hot-forming of metals such as beryllium and titanium, to provide catalytic combustion surfaces, gas turbine silencers and mufflers, high temperature gaskets and seals for expansion joints, and high temperature filtration. The higher temperature forms (to 2600°F) of alumina-silica fiber are also made into blankets. High-silica fibers and zirconia fibers are also of interest here, permitting end-use temperatures of 2400°F (when treated) and 2900°F, respectively. In addition, the zirconia fibers exhibit high chemical durability, resisting attack by hot concentrated sulfuric and phosphoric acids, cold aqueous hydrofluoric acid and its salts, and also by strong alkalis.

Ceramic fiber boards are fabricated in sizes up to 24 X 48 in. Thickness range is from $\frac{1}{2}$ to 2 in. (combinations and layering are widely utilized), and density ranges up to 16 lb/cu ft (blocks to 21 lb/cu ft). General utilization for boards is in the following fields: furnace and kiln backup insulation, thermal covering for stationary steam generators, over-the-road linings for molten metal in ladles (interplant transfer), aluminum holding furnace insulation, and cover insulation in the chemical process industry such as for magnesium cells and all high temperature reactors.

Utilizing both blankets and boards, one of the most interesting and productive applications for ceramic fibers is in linings for industrial furnaces. The lightweight ceramic fiber material may be exposed to the "inferno" of the inside of furnaces or kilns at temperatures up to 2300 or 2600°F (see Fig. 3-35).

---

*Needle-punching technique to form non-woven mats and fabrics.

Fig. 3-35A. Application of refractory ceramic fiber as lining for a high temperature furnace. Blankets are retained by special metal or ceramic studs. Temperature resistance to 2300°F is possible. (*Courtesy Johns-Manville Sales Corp.*)

Construction and installation are exceedingly simple.   A metal furnace shell is erected to size.   Studs are welded to the casing extending inward.   High temperature alloy metal or ceramic extenders are affixed, the refractory fiber blanket or board material is impaled, and quarter-twist alloy or ceramic washers are applied to contain the applied material.

Advantages are many, and include the following:[26] weight of the ceramic lining is only 25% of the weight of insulating firebrick lining, and 5 to 10% of the weight of dense refractory linings, thus permitting greatly reduced weight in furnace design; heat capacity of the ceramic fiber material is proportional to its weight, hence much less heat is required to bring the furnace to operational temperature and maintain it; there is almost complete resistance to thermal shock,

Fig. 3-35B. Close-up of impaling insulation on metal studs. (*Courtesy Johns-Manville Sales Corp.*)

so no "spalling" of refractories must be contended with, either in furnace heatup, cooldown, or door openings; no drying cycles are required and ceramic fiber-lined furnaces may be fired to top temperatures immediately without concern for fracturing due to drying or crystalline phase changes, and this adds up to longer life for the ceramic fiber lining; the low thermal conductivity of the ceramic fiber lining permits use of thinner furnace walls, simplifying construction and providing increased furnace volume, also more accurate temperature control; since the lining is not rigid, but resilient and pliable, mechanical shock damage is precluded, and furnaces can be shop-fabricated for delivery; linings are easy to install and do not require skilled tradesmen; also repairs may be easily made to damaged portions by simple cutout and replacement. Ceramic fibers may also be applied directly over new or existing refractory brickwork.[27]

In Table 3-13 is presented a comparison of use of ceramic fiber lining materials with brick-type furnace facings. A combination of

TABLE 3-13. Comparison of Ceramic Fiber with Other Furnace Linings for Equivalent Shell Temperatures.

		Heat Loss (Btu/sq ft/hr)	Heat Storage (Btu/sq ft)	Weight (lb/sq ft)
CERAMIC FIBER LINING	3″ 3″ 1800F Ceramic Fiber Lining Mineral Fiber 1303F 182F	220	1,546	5.75
INSULATING FIRE BRICK	9″ 1800F JM-20 175F	201	4,603	22.0
FIRECLAY BRICK	9″ 1800F 424F	1,239	23,400	98

Feature	Ceramic fibers	Insulating firebrick
low thermal conductivity	Better	Worse
inexpensive to buy	Worse	Better
inexpensive to install	Better	Worse
adequate refractoriness	OK	OK
long lived	OK	OK
low heat storage	Better	Worse
good spalling resistance	Better	Worse
mechanically resilient	Better	Worse
no reheat expansion or shrinkage	Better	Worse
inexpensive to repair or maintain	Better	Worse

3 in. thick ceramic fiber lining backed up with 3 in. thick high temperature mineral wool thermally outperforms a 9 in. thick wall of either insulating firebrick or fireclay refractory brick.

With any series of "pros" there are bound to be "cons," and ceramic fiber furnace linings are no exception. Lest the good reader get the impression that the heat-processing industry is carried away, it is worthwhile to point up cautionary limitations to the use of ceramic fibers as a furnace lining material:[28] ceramic fibers are more expensive, hence are less cost-effective than insulating firebrick; they are not better insulators than firebrick at 2000–2300°F, but are better at lower temperatures (1800°F), hence command attention as backup insulation; at higher temperatures of use, mineral fiber backup insulation becomes less effective and must be supplemented with additional ceramic fiber; all alumina-silica type ceramic fiber is subject to devitrification, not seriously affecting thermal performance but resulting in some brittleness; shrinkage of ceramic fiber material at top operating temperatures may open gaps at edges, requiring additional studs, crisscross lacing of studs with high temperature alloy wire to prevent sagging, or butting fiber installation to stud rows and covering with batten strips impaled on the studs; high velocities of gas or atmosphere inside furnaces may have a deleterious, degrading effect on ceramic fiber, and must be compensated for by wiring, or installation of expanded high temperature alloy metal lath if gas velocities approach 600 to 3000 ft/min.

Ceramic Fiber Papers

Processing washed, shot-free bulk ceramic fiber through a Fourdrinier machine yields a series of useful papers. Organic binder up to 5% may be added, or the fiber may be left unbonded. Fiber lengths up to 1 in. are used. Thicknesses from $\frac{1}{32}$ to $\frac{1}{8}$ in. are usual, but may be layered. Density is in the range of 12 lb/cu ft. Major applications are as speciality components in high temperature processing: gaskets; combined thermal and electrical insulation (dielectric strength = 100 V/mil); lining for combustion chambers, metal trough backups, hot tops and ingot molds (high-purity metals and glass); as a thin parting agent in metal and ceramic forming process; and others. Ceramic fiber paper may be rolled to form laminated tubes and discs of many varied dimensions, and may also be stamped or die-cut for electronic components, glass-feeder-bowl orifice rings, etc.

Ceramic Fiber Textiles

In order to provide shaped ceramic fiber components with some directional tensile strength (greater than that for batts, felts, and boards), bulk fibers (usually the long variety with shot removed) are combined with 15 to 25% organic fiber, carded, and made into strands. These may subsequently be braided into sleeving ($\frac{1}{2}$ to 4 in. I.D.), into flat tapes (3 to 6 in.) or fabrics (to 72 in. wide) braided into round or square rope of 2 in. diameter or square (see Fig. 3-36).

Although the organic fiber burns out in preliminary firing, glass fiber or metal wire may be incorporated to preserve tensile strength. Temperature limitations are: fiber glass strand—1000°F, stainless steel wire—1500°F, nichrome wire—to 2000°F. The following comprise major end uses: gasketing and wrapping insulation; coverings for induction-heating furnace coils; cable and wire insulation or braided sleeving; infrared radiation diffusers; insulation for fuel lines and high-pressure steam portable flange covers. The ropes are

Fig. 3-36. Illustration showing some of the varied forms of woven products made using ceramic refractory fibers. (*Courtesy Carborundum Co.*)

used effectively in gasketing and seals, and may be enclosed in wire mesh, or coated with graphite for increased lubricity.

Vacuum Forming Special Shapes

Shot-free bulk ceramic fibers and inorganic binders which provide some adhesive action without immediate firing are combined in a water slurry. This mix may be charged to screen-type cavity molds and a vacuum drawn to build up required thickness of the ceramic fiber mass; or a formed screen may be introduced into the slurry, a vacuum exerted, and a layer of the fiber accumulated on the outside of the screen. In either case, the part may be subsequently dried and removed by draft or taper previously designed into the form.

Plastic castable mixes of fiber plus the inorganic binder may also be formulated and applied wet or in the pliable state directly to a form to be insulated. Flat discs with flanges, short pipes, tubing, elbow bends, cones, closed-end cylinders with extensions or openings, and many other shapes may be fabricated using this vacuum casting technique. (see Fig. 3-37). Built-up thicknesses from $\frac{1}{8}$ to 2 in. are possible, and the density range is 12 to 40 lb/cu ft. As is standard for alumina-silica ceramic fibers, temperature end-use limit is 2300°F. Some shrinkage allowance must be made.

The ability to produce shapes to precise dimensions on at least one surface permits these vacuum cast parts to be widely used. Following are examples of end-use applications: in smelting, casting and foundry operations as top-out or top hole plug cones, riser sleeves, feeder tubes, hot tops, ladle linings, and dip tubes (nonferrous metals); combustion chambers for domestic heating equipment (gas or oil); high temperature pipe insulation; skid-rail insulation in steel billet reheat furnaces; thermal battery liners; heating-element support pads; and finally one of the most important—lining and insulation for automotive catalytic converters to reduce air pollution from car engine exhausts.[29]

Mixes

A series of combinations of ceramic fiber with like refractory materials has been developed. These are described individually as follows.

1. *Tamping Mixes.* Parts such as large crucible lids may be lined with a mix of ceramic fibers plus inorganic binder materials. The mix is

Fig. 3-37. Vacuum-cast refractory shapes made from ceramic refractory fibers. Shown are one-place combustion chambers for domestic oil-fired furnaces and also assorted shapes for industrial applications. (*Courtesy Johns-Manville Sales Corp. and Carborundum Co.*)

tamped in and allowed to dry. A servicable outer skin develops, protecting the inner lightweight insulating material. The 24 hr linear shrinkage is 1.7% at 1800°F and 5.0% at 2300°F. In addition to the covers, other applications include: boiler access door and furnace damper linings; linings for chemical and petroleum processing vessels; packed-on backup insulation for furnaces and kilns; muffler-pipe insulation for diesel exhaust systems; stand-pipe insulation for coke ovens. When mixed with hydraulic-setting inorganic binders or cements, castable or trowel-able mixes up to 60 lb/cu ft (dry) are formed. These are used as heavy-duty linings and backup insulation in many varied high temperature industrial and chemical processes.

2. *Composite Insulation for Space Firings and Launchings.* A NASA requirement is described for launch pad insulation and space vehicle protection with higher mechanical and physical strengths. This was solved by development of a mixture of alumina-silica ceramic fibers, pigmentory potassium titanate, and asbestos fibers.[30] Stresses, strains, shocks, and vibration patterns of a firing rocket engine were easily handled, as well as both radiative and convective heat.

3. *Reinforcement of Zirconia and Like Foams.* Zirconium oxide, mullite, and calcium aluminate foams are proposed and show good resistance to direct exposure to a rocket flame at up to 3000°F. Physical conditions were: heat flux—40 Btu/sq ft, vibration—60 cps, $\frac{1}{2}$ in. double amplitude displacement and 90g acceleration. Ceramic fibers were evaluated as a refractory strengthener.[31]

FILTRATION

Introduction

Considering macro and micro worlds, an old poem comes to mind:

Title: "∞"

Big fleas have little fleas
With littler fleas that bite 'em,
And these same fleas have smaller fleas
And so ad infinitum.

Despite all the mightiest efforts to resist temptation, the following bit of terse verse reared itself to correlate the similarity of modern fiber glass filtration materials to "∞" above. (Quotes a hypothetical conversation between filter media and particles attempting to pass through.)

Where're you going, little speck?
Fibers of glass will grab you by the neck!
Perchance your smaller chums go through,
Minuter fibers will snag them too!

So, if through a filter you propel,
There's no choice but to ring your bell!
You should go home and tell your pappy
We must keep EPA and OSHA real happy.

Filtration problems generally resolve themselves into two main classifications: air (gases) and liquids. Cleaning and purification technology has come a long way toward perfection since early use of glue-bonded horsehair for air filtration or a chamois stuffed in a funnel for filtering engine fuel. In this section are described the versatile forms and superior behavior of fiber glass as a filter medium in both these areas.

Air Filtration

Condition of Air Requiring Filtration

A portion of a history of fiber glass prepared by the Owens-Corning Fiberglas Corporation, describing the first successful commercial vitreous filamentary product of the modern era, reads as follows:

"John H. Thomas and a group of associates began the study of glass fiber production under the direction of the late Dr. Games Slayter. A small research laboratory was established in the Owens-Illinois plant in Evansville, Indiana, where coarse glass fibers were produced on a pilot-plant basis. These fibers were immediately developed into an air filter that was marketed in February 1932—less than nine months after the first attack at the problem. This first Fiberglas product became the Dust-Stop® Filter which is now standard equipment in forced warm air heating units and is distributed widely for filtration in air conditioning equipment."

In Figure 3-38, Part A is presented a photo of an automated paint-line utilizing these early filters, and in Part B a close-up view of their appearance as then manufactured.

The "gremlins" that lurk as particles carried through the air are notably undesirable because, different from most of their counter-parts of a polluting nature in water, etc., they cannot always be seen or easily detected. Yet they are there as a result of heavier industrial-

Fig. 3-38A. An automated paint line using air filters soon after their development in 1932. (*Courtesy Owens-Corning Fiberglas Corp.*)

Fig. 3-38B. A close-up view of characteristic early types of air-cleaner filters. (*Courtesy Owens-Corning Fiberglas Corp.*)

ization, increasing travel by automobile, and yes—even continued and increased smoking by members of the human race.

The particles suspended in a gas, particularly air, comprising smokes, mists, fumes, grits, dusts, and other dry granular particles, lint and fibers, are referred to as "aerosols." The degree to which they exist and their particular concentrations in the air all around us is critical, and increasing attention is being given to methods for their removal, especially from air brought into buildings in which people live and work.

After analyzing air samples taken from different areas of Great Britain,[32] one researcher concluded that pollutants vary, that contamination comes from many sources, and that the quantity will vary significantly with time of day or year, and prevailing weather conditions. Hence, air conditioning and ventilating facilities must answer many varying requirements.

In rural areas in Great Britain, particle concentration in air was found to be 0.05 to 0.5 mg/m^3, and consisted of soil from wind erosion, vegetable matter, carbonaceous material, and pollens. In coastal areas, concentration was almost identical, except that salt particles were also present.

In metropolitan areas, dust concentration increased to 0.1 to 1.0 mg/m^3, and contained increased amounts of carbonaceous matter 0.5 to 2 μ in particle size, probably introduced from incomplete combustion of solid and liquid fuels. The metropolitan air samples also contained ash composed of SiO_2 and other granular particles, and included erosion products from roadways and construction sites.

In industrial areas, naturally the worst, concentration of air pollutants ranged from 2 to 5 mg/m^3. The type of contaminants would depend upon local production activity, but included fine carbonaceous materials, tarry oils, mineral and chemical products, and sulfurous gases and acids.

If airports and highways were nearby, tars and waxes increased in concentration and were found in particle sizes up to 2 μ. It was found that large micron-size particles fall to the ground very close to their point of entry. It is well known that a high-pressure blast from a 120 psi air hose will move rocks, but particles much larger than 50 μ do not remain airborne under normal atmospheric conditions.

In another investigation,[33] complete size and quantity analysis was given of an air sample taken by the University of Minnesota. This is duplicated in Table 3-14.

TABLE 3-14. Complete Analysis of Sample of Atmospheric Dust.

Average particle size (μ)	Range of particle size (μ)	Proportionate quantity by particle count	Percent by volume (or wt. if uniform sp gr)
20	10–30	1,000	28
7.5	5–10	35,000	52
4	3–5	50,000	11
2	1–3	214,000	6
0.75	1–1.5	1,352,000	2
0.25	0–0.5	18,280,000	1

Additionally, considerable research and investigation have been performed on the actual particle sizes and ranges of known substances encountered as industrial or associated air pollutants.[34] These are shown in Fig. 3-39. Note the extremely wide range encompassed in micron particle sizes.

Properties of Glass Fiber as an Air Filter Medium

Well, the question of what to do about containing all these culprit materials is becoming increasingly more critical. In the early stages of development of the technology of air conditioning, air could be mechanically distributed at controlled temperature and humidity, but properly cleaning it at the same time became long overdue. Fortunately, air cleaning means and methods are now receiving a justifiable share of attention.

Hence, enter the science of filtration! Whereas persons in homes, hospitals, offices, and stores "existed" in the last generations, their lives have been made much safer, easier, and more pleasant, and work outputs substantially increased by improvements in filtration technology. As an example, it was found that 90% of the staining of wall decorations, light fixtures, the "hum" on window lights, and discoloration of goods displayed in stores on exposure to room air was caused by aerosols (airborne particles) 1 μ and less in diameter. Many other problems existed such as the need for greater air cleanliness in hospital operating rooms and elimination of contaminants in nuclear, optical, and space-age laboratories and associated production operations. These hastened improvements in filtration procedures by demanding greater sophistication.

Whereas originally many materials were used for filtration including

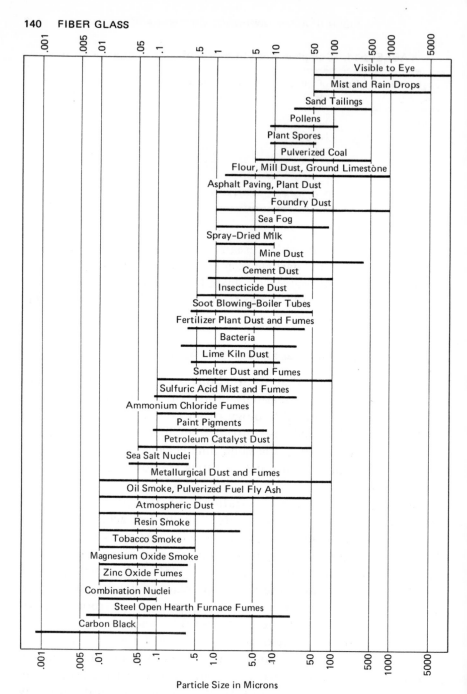

Fig. 3-39. Micron particle size and ranges of various domestic and industrial air pollutants. (*Courtesy Johns-Manville Sales Corp.*)

natural fibers, plastic-covered or bonded horsehair, and many others, newer materials such as glass fibers, synthetic fibers, expanded metals, and even electrostatic precipitation proved their effectiveness and came to be used.

Of all the various materials available for air filtration, glass fibers have become the most desirable and effective because of their unique combination of properties, itemized in part in the following: fibers of glass are strong, durable, chemically resistant and essentially inert; they are nonhygroscopic, and hence free from bloating or swelling; they will not fibrillate nor gelatinize as will cellulosic and some other synthetic fibers; their high temperature thermophysical constants permit their use at points well above room temperatures, and hence they present no fire hazard; they exist in rodlike form, the most desirable shape for a filter material; they may be process-controlled to yield a wide span of fiber diameters in narrow-ranged groups, thus providing sensitivity for selective removal of particles in different graded micron-size classes; fiber glass may be formed into webs, mats, blankets, papers, membranes, or other forms of filter media, each having a high percentage of voids, thus providing large dirt-holding areas; fiber glass mats, blankets, etc., may be shaped into the various elements required by the specific filtering problem or condition, i.e., panels, folded structures, sleeves, bags, etc.; fiber glass may be prepared with or without binder, with binder either cured or uncured (for later processing), and it may be shipped collapsed under vacuum in sealed rolls for considerable space saving.

Adaptation of fiber glass to air filtration and to similar applications has brought about creation of two separate industry coding systems. These are listed successively in Tables 3-15 and 3-16. The materials listed in Table 3-15 are blanket types developed essentially for air filtration, while Table 3-16 presents the code designation for Micro-Fibers® and Micro-Quartz® from which specialty papers and felts are produced. In addition to the parameters indicated in Table 3-15, glass fiber filtration media are also tested for ignition loss (or binder content), resiliency, and recovery after loading with contaminant.

In addition to the above, several other types of specialty glass fiber filter elements have been developed. These include graduated-density mats and membrane prefilters which will receive consideration later under Liquid Filtration.

TABLE 3-15. Types and Coding of Fiber Glass Air Filter Blankets.

Manufacturers' designations	Average filament diameter*	Thickness (in.)	Color designation	Weight (g/sq ft)	Tensile str lb/in. width, mach./x-mach. direction	Avg resistance to air flow, in. of water column (wc)	ASHRAE dust-spot filter eff %
No. 12 (AAF) Bonded mats	U G–K	Up to 2 To 0.030	Blue White or brown	As required	—	0.05 —	35 —
HEAF	B	To 0.75	Yellow	—	—	—	—
General purpose	B	0.25, 0.50	Yellow	9.5, 18.9	—	0.05	35
AF-21 (JM)	B (0.00016 in., >3.8 μ)	0.50 ± .06	Yellow to light tan	11.4 ± 1.4	5.9/1.4	0.06 ± .02	30
AF-18 (JM) FM-018 (OCF)	B (0.00016 in., >3.8 μ)	0.27 ± .05	Yellow to light tan	9.5 ± 1.1	3.0/0.75	0.07 ± .02	30
AF-11 (JM)	0.00011 in.	0.27 ± .05	Orange	5.7 ± 0.7	1.0/0.40	0.07	55
FM-11 (OCF)	0.00011 in.	0.25, 0.50	Orange	—	—	.05 .085	55
AF-4 (JM)	.00004 in.	0.25 ± .03, 0.50	Pink	4.3 ± 0.5	0.45/0.20	0.21 ± 0.03	85
FM-004 (OCF)	.00004 in.	0.31	Yellow	—	—	0.55	85
AF-3 (JM) FM-003 (OCF)	.000003 in.	0.27 ± .05	Yellow	4.8 ± 0.6	0.45/0.20	0.46 ± .06	95

AAF = American Air Filter Co./JM = Johns Manville Sales Corp.
OCF = Owens-Corning Fiberglas Corp.
*See Tables 3-2, 3-16 and 5-2.

TABLE 3-16. Types and Coding of Ultrafine Glass.

Industry code	J-M code	Fiber diameter Avg (μ)	Fiber diameter Avg (in.)
B	Code 112	2.6–3.8	.00010–15
A	Code 110	1.6–2.59	.00006–10
AA	Code 108	0.75–1.59	.00003–06
AAA	Code 106	0.5–0.749	.00002–03
AAAA	Code 104	0.2–0.499	.000008–20
	Code 102	0.1–0.199	.000005–08
	Code 100	0.05–0.099	.000002–05

Understanding Air-Filtration Technology

As pointed out above (Table 3-15), filter materials are judged by efficiency. Pressure-drop in inches of water column was also alluded to. A third parameter, arrestance, must be defined, and a fourth requirement of a filtration medium (or media), dust-holding capacity, is obviously of importance. Other elements in determining performance of an air filter are fiber diameter and fiber packing area and thickness of the media, and velocity of air flow through the media. Also, sizes, shape, specific gravity, and concentration of the aerosols constitute other variables.

Prior to 1960, considerable ambiguity existed for definitions of filter efficiency and performance in original test methods generated by the many different agencies in the U.S. and abroad. Many of these had "conditional" or "judgment" factors that led to natural errors, spurious interpretations, and general "fudging" of results. Without belaboring the reader with needless and cumbersome explanations, it is sufficient to report that the American Society of Heating, Refrigeration, and Air Conditioning Engineers (ASHRAE) consolidated three of the four main "filtered-out quantity" parameters and incorporated and gave credence to a fourth. As a result, an across-the-board, complete performance evaluation of a filter media can now be determined and reported.[35]   Also, under supervision of ASHRAE, a standard testing unit was developed which gave reliable and duplicative results for the following parameters.

1. *Arrestance.* Coarse filtration is needed to separate larger particles present in normal to excessive quantities. ASHRAE also developed a synthetic dust for evaluation of arrestance. It is composed of

72% (by weight) standardized aircleaner dust (fine); 23% Molocco Black; 5% No. 7 cotton linters ground in a Wiley mill, 4 mm screen. Sources of these materials plus method of fabricating the test dust are given in the ASHRAE literature. Results are reported as percent weight arrestance to differentiate from:

2. *Atmospheric Dust-Spot Efficiency.* This test determines the quantity of smaller particles 5 μ or less in diameter which are removed by any given filter media. The determination is made by observing actual discoloration due to the contaminants of a white filter paper (Hollingsworth and Voss H-93), targets interposed upstream and downstream from the test sample. Unaltered atmospheric air is the only input material. This test evaluates the ability of the filter media to remove particles which would cause discoloration of walls and other interior furnishings or goods. Results are reported as percent atmospheric dust-spot efficiency, and this represents the percentage evaluation figure normally referred to unless otherwise designated.

3. *DOP Efficiency or Penetration.* To measure performance of ultrafine filters, a vaporized fume or smoke from di-octyl phthalate (80 mg/m^3) is directed into the test media, and the amount passing through is determined photoelectrically. Penetration is usually specified rather than efficiency because such high-performance filters have efficiencies near 100%. (The procedure for conducting the DOP efficiency test is outlined in MIL-STD 282.)

4. *Dust-Holding Capacity.* The standard ASHRAE test dust is used, and charged to the test sample at least four times. Dust-holding capacity is enhanced by high permeability or increased (extended) surface area. Results are reported in g/1000 cfm cell.

Other pertinent factors such as pressure drop and leakage are also determinable in the ASHRAE test. Pressure drop (ΔP) is the result of resistance to air flow, and is defined as change in the air force when a filter is placed in an airstream. ΔP is measurable in the ASHRAE test, and is expressed as inches of water column (WC). All filter media are rated for air flow, hence rate in feet per minute must be specified for any ΔP data.

The important thing to be realized is that the three major different tests, arrestance, dust-spot efficiency, and DOP efficiency, must be conducted to evaluate the true performance of any particular air filtration media, and the method of test should be well detailed in any data. Obviously, the term "efficiency" should not be used loosely or without qualification. Table 3-17 shows comparative data on these three methods for glass fiber and several competitive filter media materials.

It is of interest here to study how air filters actually function in re-

TABLE 3-17. Performance of Dry Media Filters.

Filter media type	ASHRAE weight arrestance (%)	ASHRAE atmospheric dust-spot efficiency (%)	MIL-STD 282 DOP efficiency (%)	ASHRAE dust-holding capacity (g/1000 cfm cell)
Finer open cell foams and textile denier non-woven	70–80	15–30	0	180–425
Thin paperlike mats of glass fibers, cellulose	80–90	20–35	0	90–180
Mats of glass fiber, multi-ply cellulose wool felt	85–90	25–40	5–10	90–180
Mats—5 to 10 μ fibers ¼–½ in. thickness	90–95	40–60	15–25	270–540
Mats—3 to 5 μ fibers ¼–¾ in. thickness	>95	60–80	35–40	180–450
Mats—1 to 4 μ fibers—mixture of various fibers and asbestos	>95	80–90	50–55	180–360
Mats—½ to 2 μ fibers (usually glass fibers)	NA*	90–98	75–90	90–270
Wet laid papers of mostly submicron glass and asbestos fibers (HEPA filters)	NA	NA	95–99.999	500–1000
Membrane filters (membranes of cellulose, acetate, nylon etc. with micron-size holes)	NA	NA	100	NA

*NA indicates that test method cannot be applied to this level of filter. (*Courtesy ASHRAE.*)

moving particulates. Considering a filamentary, random-mass material, exemplified by fiber glass, the voids created or defined by fibrous strands or groupings may be either smaller, approximately equal to, or vastly larger than the particles to be removed from the air. An airstream breaks up into many smaller currents as it moves in a circuitous path through a fibrous filter mass. Filtration takes place when a dust particle contacts and adheres to one of the fibers, or to a previously deposited foreign body.

Deposition of a particle on a fiber occurs in one of three ways.

1. *Impingement* (*Impaction*) is usually involved when large particles with comparatively greater mass are traveling at high velocity, move into a filter medium, and encounter a fiber. Having some inertia, the larger particles are more prone to resist change in direction, hence continue essentially on course, soon striking a fiber. A viscous oil or adhesive may be applied to filters for coarse particles (impingement type) to assist in holding the particles once they meet the fibers. Throughout the usage period or life, adhesive-coated filters exhibit higher percentage of arrestance than dry filters without significant differences in increase in resistance to air flow.

2. *Interception* (*Diffusion*). An interception-type filter consists of an extremely fine fibrous mass. As the small dust particles borne by an airstream move into the body of the filter media, the number of fibers encountered is very large. Hence, the probability is almost nonexistent that a dust particle will not contact a fiber. The principle of Van der Waals forces is responsible for the bond, effectively securing the dust to the fiber.

3. *Straining.* This principle comes into play when relative sizes of dust particle and opening favor it, and also when sufficient dust particles have accumulated to help deter passage of newly entering particles. Obviously, a very fine-fibered filter is required.

Types of Filter Equipment, Corresponding Fiber Glass Media, and Characteristic Applications

The type of equipment developed to harness incoming air and direct it through the required media is both ingenious and interesting. Correlating equipment with the type fiber glass media and the typical end applications is the purpose here.

Each piece of equipment consists mainly of a housing to hold the filter media and channel the air, and also a fan or other means to force the air through the media. There are four major types of equipment and several miscellaneous types, described according to atmospheric dust-spot efficiency rating as follows.

1. *Low Efficiency (15–30%).* Referred to as the "roughing" filter, typical equipment is that which holds panels (Fig. 3-38), and units in which fiber glass blankets are either automatically or manually un-

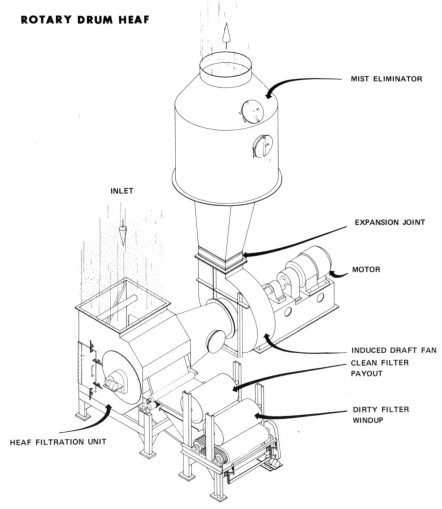

ROTARY DRUM HEAF

MIST ELIMINATOR

INLET

EXPANSION JOINT

MOTOR

INDUCED DRAFT FAN
CLEAN FILTER
PAYOUT

DIRTY FILTER
WINDUP

HEAF FILTRATION UNIT

Fig. 3-40A. An isometric drawing of the component operable parts of a HEAF (High Energy Air Filtration) unit. The term "high energy" refers to the accelerated speed at which the particles are driven onto the media, usually greater than 500 ft/min. (*Courtesy Johns-Manville Sales Corp.*)

rolled so that a single layer of the media intercepts an airstream, removing coarse particulates. Such a unit is the HEAF (high energy air filtration) shown both schematically and in actual operation in Fig. 3-40. Fig. 3-41 shows SEM photomicrographs in cross-sectional aspect of HEAF filter media both before and after use.

Fiber diameters of fiber glass in the panel-type filters range from 10 to 12 μ in the blown-fiber type, to 0.001 in. for the drum-wound type. Diameters for filaments in the HEAF and similar type "roughing" filter blankets are in the 3–4 μ range.

Typical applications for panel filters are as media for dust separation out of forced-draft furnaces and air conditioners for homes and commercial buildings. They are also used as prefilters to protect

Fig. 3-40B. An actual HEAF installation. The continuous feed of fresh fiber glass blanket material and the roll-up of the used, exhausted stock are readily visible at the right-hand portion of the illustration. Unit manufactured by Anderson 2000 Inc. U.S. Patent Number 3,745,748. (*Courtesy Johns-Manville Sales Corp.*)

Fig. 3-41. SEM photomicrographs at 100X of the cross section of HEAF filter media both before and after use in an operation removing phenolic resin particles from exhaust gases from a fiber glass operation producing bonded industrial mat. (*Courtesy Johns-Manville Sales Corp.*)

medium-efficiency filters. Viscous oils or adhesives are added to increase dirt-gathering capacity.

Units such as the media-throughput HEAF are used to separate out coarse and undesirable or illegal contaminants from air or stack exhaust in industrial processes. Filters of this caliber will also remove lint, ragweed, and some other pollens.

2. *Medium Efficiency (30–85%)*. Extended filter surface is the key to success in this type filter equipment. Frames are designed for incorporation into a main housing. These frames hold and support the media which has been fluted, pleated, made into bags, or otherwise fabricated to increase surface area. To illustrate behavior, an airstream of up to 625 ft/min velocity can be reduced to less than 100 ft/min by intercepting with a highly pleated filter. Extending surface by pleating, bagging, etc., as described, has also been found to provide an almost undetectable increase in ΔP for 75% of the effective filter life. In Fig.3-42 is illustrated a schematic drawing of the frame plus pleated and bagged filter element, and an actual view of the exit-air end of such a pleated filter in a large installation.

Filament diameters for glass fibers capable of performing in this efficiency range vary from 0.00016 in. down to 0.00004 in. (3.8 to 1.0 μ). Particles filtered out are essentially below 5 μ.

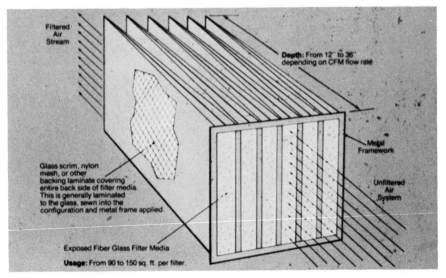

Fig. 3-42A. A schematic drawing of holding frame and assembly of pleated fiber glass filter media. (*Courtesy Johns-Manville Sales Corp. and Cambridge Filter Co.*)

Fig. 3-42B. The exit end of these pleated segments in an actual installation. (*Courtesy Johns-Manville Sales Corp. and Cambridge Filter Co.*)

Typical applications are central domestic heating and air conditioning systems, for make-up air entering paint-spray systems, and as prefilters to the high-efficiency class.

3. *High Efficiency (85–95%)*. Systems identical to those for the medium-efficiency range (frames to hold pleated, bagged, fluted, etc. media) are used. The exception is that the fiber glass media used has a nominal diameter of 0.000003 in. (0.075 μ).

Particles filtered out are those 1 μ and below, which account for staining. This class also removes all pollens and particles, smudge and fumes produced by coal, oil, and tobacco smoke. Bacteria are also entrapped. The filtration capability of this class is required in hospital operating rooms, areas for production of pharmaceuticals, data processing, nuclear and aerospace laboratories and production areas.

4. *Absolute (HEPA) Filter (99.97%)*. Micro-Fibers® remove 99.97% of 0.03 micron-sized particles. Fiber glass forms used in this application comprise the code 106, 104 and 102 (Table 3-16). The three types have specific functions for removing similar but slightly different particle types in this exceedingly fine range. The microfiber is formed into thin paperlike sheets which are supported by metal screening and/or framing for adequate support and to prevent leakage. Electrostatic precipitators are sometimes used to accomplish air cleaning and remove particles in the 0.03 μ range. Operation and

maintenance of these electrostatic units are extremely costly. Mechanical cleaning is required, during which time the unit is usually out of operation. While in operation, dust accumulation frequently causes short circuits.

By contrast, fiber glass media is always of the throwaway class, either intermittent or continuous, automatic replacement. The pressure drop, ΔP, is continuously monitored using an inclined manometer. As soon as 1 in. water column or greater is reached, the filter life is ended and easy, rapid replacement is made. The anomaly surrounding increase of efficiency of a filter media with use, like natural phenomena (nothing is free in nature), is not a free ride. It must be paid for by an increase in power required to drive the fan and maintain some semblance of the required air flow rate through the filter.

5. *Miscellaneous and Variations.* Blanket filter media will usually have a backing of loose scrim, open-weave fabric, Cerex, or a metal for support and to prevent rupture and consequent leakage. Filter life may be extended and energy conserved by use of the Variable Air Volume (VAV) Concept (see Pipe and Duct Insulation). Instrumentation systems control the conditioned space temperature in an air conditioning or heating installation by automatically reducing the air flow whenever the actual load is less than specified design conditions. Variance from 20 to 100% may be allowed for. With this VAV system, full advantage may be taken of diversity of load due to occupancy levels, solar heat conditions, and lighting levels. This system requires higher than normal filtration efficiencies (atmospheric dust spot). Electrostatic air cleaners have been generally acceptable for this system.

Another energy saver proposed is the reverse-cycle system, which may be described as closed-loop heat recovery. This system is not regarded as fully acceptable. It does not employ a central ducting system. Filtration standards are poor because the disposable media cell is only $\frac{1}{2}$ in. thick.

Performance Data for Fiber Glass Filter Media

Data on variation of ΔP with fiber diameter, air velocity, media density, and dirt load accumulated has been developed and is presented in Fig. 3-43. These will further assist the reader in understanding and interpreting the values of fiber glass as an air filter media.

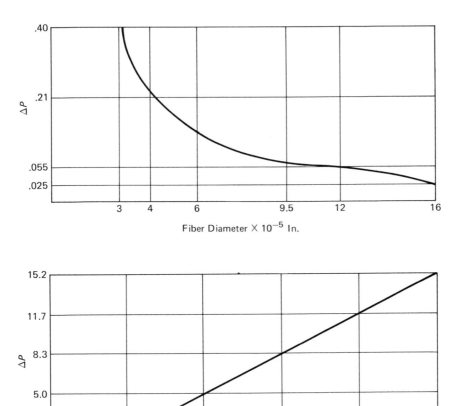

Fig. 3-43A. General curves showing change in ΔP for various properties and conditions of glass fiber filter media. (*Courtesy Johns-Manville Sales Corp.*)

Filtration of Liquids

Introduction

As in air filtration, a sizable, well-developed technology has also grown around removal of undesirables from liquid systems. Examples are juices and beverages, foods, chemicals, pharmaceuticals, liquor, inks, plastics (latexes, etc.), jet fuel, photographic processing solutions, swimming pool water, and many others.

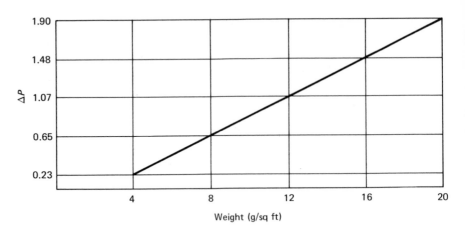

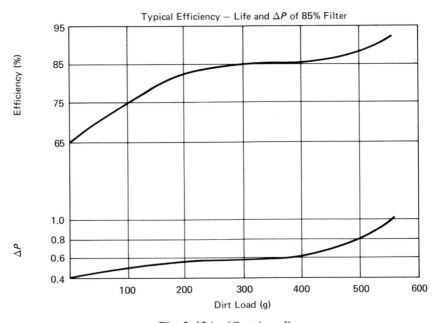

Fig. 3-43A. (*Continued*)

Assayed here is the role of fiber glass as a material capable of wide-spectrum performance in this area. Basic rationality behind these technical advances is explained. Construction and performance of fiber glass filters and discussion of several specific applications round out this section.

Specification Data:

ASHRAE Test Data on Various Typical Filter Configurations

Efficiency vs. Life

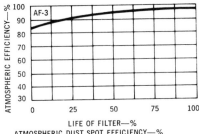

LIFE OF FILTER—%
ATMOSPHERIC DUST SPOT EFFICIENCY—%
INITIAL: 82-86; AVERAGE: 90-94; FINAL: 95-98 +

Pressure Drop vs. Dust Load

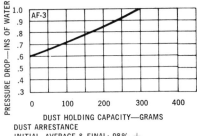

DUST HOLDING CAPACITY—GRAMS
DUST ARRESTANCE
INITIAL, AVERAGE & FINAL: 98% +

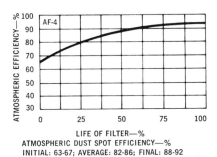

LIFE OF FILTER—%
ATMOSPHERIC DUST SPOT EFFICIENCY—%
INITIAL: 63-67; AVERAGE: 82-86; FINAL: 88-92

PRESSURE DROP VS. DUST LOAD

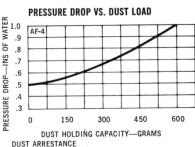

DUST HOLDING CAPACITY—GRAMS
DUST ARRESTANCE
INITIAL, AVERAGE & FINAL: 98% +

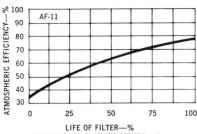

LIFE OF FILTER—%
ATMOSPHERIC DUST SPOT EFFICIENCY—%
INITIAL: 33-37; AVERAGE: 54-58; FINAL: 78-82

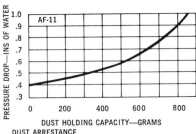

DUST HOLDING CAPACITY—GRAMS
DUST ARRESTANCE
INITIAL: 66-70; AVERAGE: 84-88; FINAL 94-98

The data shown on these graphs is representative of the average performance of various filter configurations and designs currently being used by filter manufacturers. The filters were constructed of approximately 100 square feet of   filter media. The air flow rate was 2000 CFM. Specific values will vary somewhat depending on the filter design. The test method was ASHRAE Standard 52-68, "Air Cleaning Devices" 8/16/68.

Fig. 3-43B. Curves showing efficiency vs filter life and pressure drop vs dust load for three specific fiber diameters of fiber glass filter media. (*Courtesy Johns-Manville Sales Corp.*)

Filter Fabrication

Fiber glass material for conversion into the usable form for liquid filtration, the "filter tube," is a blown fiber collected on a moving belt, and mandrel-wrapped prior to curing in a manner described previously (see Pipe Insulation).

The finished filter tubes consist of four major components: (1) perforated center-reinforcing core, to resist the higher external pressures attainable with fiber glass filters, and comprising either plastic, tin-coated steel, stainless steel, resin-impregnated paper, or no core at all, depending upon the chemical solution and pressure to which the filter is exposed; (2) a thin organic fiber veil mat to protect inner surface integrity of the filter media; (3) the tubular fiber glass material in slightly graded density—higher inside to lower outside; and (4) an external fabric sleeve to preserve outer surface integrity.[36]

Several general variables may be introduced. The fiber diameter may be varied for selective micron-size filtering (fully discussed later). The external portion of the filter tube may be grooved to increase surface area. End caps may be applied for stability and to satisfy some filter-housing design requirements. In Fig. 3-44 Part A is presented a cutaway drawing showing construction and components

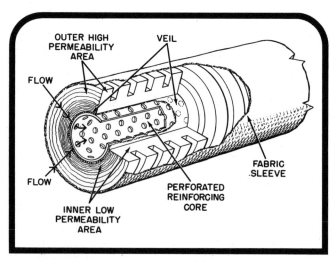

Fig. 3-44A. Cutaway drawing illustrates construction and components of a typical fiber glass filter tube cartridge.

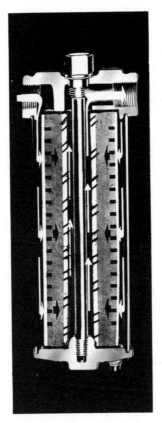

Fig. 3-44B.   The main elements and function of the housing constructed to hold the filter tube during its operation. Please note that liquid flow is from the outside of the filter element toward the inner core.

of a typical filter tube cartridge. Figure 3-44 Part B shows construction of a typical filter tube housing.

In addition to the filter tubes manufactured and used in appropriately sized housings, bulk fiber glass material is also purveyed for use in many different forms for filtration of liquids. These forms consist of bags, panels, bulk, and others.

There are several dimensional and related variables which must be defined for full understanding of filter tube application and performance.

1. *Size of Inner Diameter.* Filters are primarily selected based on inner diameter, which may vary from 0.625 in. to over 30 in. Tight

tolerances of $+0.025$ in., -0.005 in. are maintained. Sizes larger than 30 in. I.D. may be specially produced. Incremental increases in diameter supplied are approximately $\frac{1}{4}$ in. or less in the smaller sizes and 1 in. in the larger sizes.

2. *Length.* The standard replaceable filter tube cartridge is $1\frac{1}{16}$ in. I.D. $\times 2\frac{3}{4}$ in. O.D. $\times 9\frac{3}{4}$ in. long. Lengths of 36 and 72 in. are standard. Any special length requirements may be readily accommodated by the manufacturers.

3. *Wall Thickness.* The body of the fiber glass media may be varied from $\frac{1}{2}$ to 2 in. in thickness. Lesser thicknesses to $\frac{1}{4}$ in. are also produced.

4. *Densities and Interleafing.* The nominal as-produced density of the fiber glass liquid filter may be varied from 2 to 12 lb/cu ft. A natural variation occurs in that, in any specific tube type, the inner portion of the fiber glass layer may be produced at slightly higher density than the outer portion. This condition is beneficial to added filter life because the foreign material being removed from the liquid is permitted to enter the fibrous structure. There it contacts many fiber bodies, and hence does not accumulate on the outer surface only, where it would cause premature blockage.

The process of interleafing is regarded as special treatment, and consists of including auxiliary fiber glass components such as thin mats, webs, blankets, screening, or cloth. These supporting materials serve various functions from increasing mechanical strength and extending filter tube life, to making possible removal of specific contaminants.

5. *Binder Content.* Again, phenol-formaldehyde resin is used to bond the fibrous mass. A level of $9 \pm 3\%$ is considered standard. The amount may be increased to $18 \pm 5\%$ to gain mechanical strength, but other parameters such as dirt-holding capacity will be deleteriously affected.

6. *Grooving.* As stated, grooves may be machined in from the outside periphery of the filter tube to provide extended filter surface area. The minimum wall thickness of $\frac{3}{4}$ in. is required before a tube can be grooved. Groove depths are usually $\frac{7}{16}$ in. deep, $\frac{1}{8}$ in. wide, and are placed on $\frac{1}{2}$ in. centers. For purposes of stability, a $\frac{5}{8}$ in. ungrooved ring or tip is left at each end of a filter length or prepared filter cartridge.

TABLE 3-18.   Fiber Diameter vs Approximate Particle Size Removed by Fiber Glass Media (Filter Tubes) in Liquid Filtration.

Fiber glass cartridge designation[a]	APPROXIMATE NOMINAL PARTICLE SIZE FILTERED OUT OF LIQUID[b] (μ)	
	Water viscosity	Paint viscosity
C-7	75 to 100	75
C-6	50	50
C-5	25	25
C-4	15	10
C-3.5	10	5
C-3	5	3
C-2	3	1
C-1	1	½

[a]Designation of Johns-Manville Sales Corp., Filtration and Minerals Division.
[b]Size removal claimed by manufacturers of competitive non-fiber glass filters.

7. *Fiber Diameter.* The range of micron (μ) particle sizes which fiber glass and competitive filters are called upon to remove from liquids spans a slightly narrower band than that for air filtration media and airborne particles. Whereas liquid filtration material is designed to remove particles from 100 μ to slightly less than 1 μ, air filtration media will remove airborne sizes from 5000 to 0.03 μ (see Fig. 3-39).

Fiber glass fiber diameter sizes AA (0.00003 in.) to D-fiber (0.00020 in.) are used in combinations, layering, varying densities, etc., to form a series of filter media that span the 100 to < 1 μ range.

The relationship between fiber diameter and micron particle size filtered out is subject to external variables such as pressure from clamping in a filter housing, viscosity of carrier liquid, and others. However, the generally accepted averages or reference points for these ratios are listed in Table 3-18. For purposes of reference, following are micron particle diameters or sizes of some familiar substances: (1) 75 to 100 μ—grains of table salt; (2) 50 to 70 μ—diameter of human hair; (3) 40 to 50 μ—lower limit of human vision; (4) 25 μ—upper limit of optical clarity; (5) 5 to 10 μ—particle of talc (face powder).

Advantages of Fiber Glass in Filtration of Liquids

1. Fiber glass is inert and possesses high temperature resistance. Hence glass filters may be used in many corrosive environments and at temperatures to 250°F and higher. This performance is superior to that for competitive materials, usually synthetic and/or cellulosic (organic) fibers which are limited and must be used at temperatures below 125°F.

2. The fabricated filter media comprises a hard, porous structure that is strong and not susceptible to collapse or mechanical failure. In filter tube elements, selected cores may be incorporated which further contribute to strength, performance, and long life.

3. Fiber glass filters may be readily made in excessively greater lengths than can competitive materials—up to 72 in.

4. Filament diameters and, hence, pore sizes may be varied so that a wide range of specific-sized particles can be removed.

5. The clear-cut combinations of construction, fiber diameter size, tube dimensions, etc., do not make for an excessively unwieldy inventory problem. Furthermore, special sizes may be easily and rapidly produced.

6. The strong phenolic-fiber glass bond resists channeling, rupture, distortion, and reduces media migration, permitting higher ΔP levels.

7. The basic method of winding provides a graded density throughout the media, i.e., higher toward the core and lower at the outer diameter, thus creating permeability that causes entrapment of foreign particles within the body, and not merely on the surface of the filter. This also permits handling liquids of high viscosity.

8. With glass filters, less so than with other types, there is a minimal problem with extractables (fluids that might possibly react with the media, adding taste, color, or chemical contamination to the filtrate). If a questionable condition exists, filters may be preflushed or presoaked to purge extractables.

9. Depending upon end use, and foreign particles removed, many classes of fiber glass filters may be back-washed and replaced into service. Special housings and fitments have been developed for this purpose.

Testing Liquid Filtration Media

Several factors determine selection and application of filtration media:

1. *Degree or Fineness Required.* Finer filtration is more costly. The main purpose of liquid filtration is protection of equipment or an industrial operation, so the size range of particles to be removed is the primary consideration.

2. *Amount of Material To Be Removed and at What Rate.* This is sometimes referred to as the "dirt load," and determines quantity of filter elements required. Sometimes prefiltering is helpful in sharing the filter load between two different media, and extending the life of the finer filter.

3. *Overall Cost.* The overall cost of the proposed equipment, its maintenance, and filter usage rate must be balanced against their actual economic contribution.

Definitions helpful in understanding liquid filter performance are similar to those for air filtration:

1. *Efficiency* is the percentage ratio of test material passed versus that charged. It can be measured as weight percent, number percent, or by comparative turbidity of base and effluent test liquids. "Efficiency" is also employed to describe the performance of a filter, i.e., fine fiber = much material trapped rapidly = high efficiency; less material trapped = lower efficiency.

2. *Flow Rate* is the amount or quantity of contaminant-laden liquid of a given viscosity passable through the filter or filter media per unit of time. This parameter is usually expressed in gallons per minute. Flow rate for a filter cartridge system in service depends on filament diameter, number of cartridges in series, viscosity of liquid, dirt loading, and may be as high as 160 gal./min. Flow rate through a $9\frac{3}{4}$ in. filter cartridge under test is 3 gal./min.

3. *Pressure Drop* (ΔP), expressed in pounds per square inch, indicates resistance of the filter to fluid flow, and is the prime indicator of filter life. Although not recommended, some filters in service are taken to a ΔP as high as 100 psi. Some filter tube core materials resist high ΔP better than others.

Both filtration efficiency and cartridge life are functions of flow rate, and both increase as flow rate is reduced. Excessively high flow rates should be avoided by proper system design to prevent filter damage. Single 10 in. cartridges or tubes may best be replaced when ΔP = 30 psi or greater (maximum test pressure).

In testing fiber glass filter cartridges for performance, the following procedure involving turbidimetry has been adopted as standard by several large filter manufacturers' testing labs and trade organizations. It is regarded as highly reliable and duplicative as a test method. Gravimetric or particle counting methods may also be used to determine efficiency. However, the method outlined using turbidimetry is easier and more rapid in execution, and provides completely satisfactory results.

Equipment and Material Tanks or sumps for base and effluent water-dust dispersions; a single cartridge filter housing between tanks and a pump to induce metered flow through the filter cartridge; upstream and downstream pressure gages reading in psi; turbidimetry equipment (HACH surface-scatter type or equivalent); and AC* test dust.[37]

The standard test dust is known as Arizona road dirt and is supplied in "coarse" and "fine" grades. Respective compositions are indicated in Table 3-19. Coarse dust is used in almost all testing except that for filters below 5 μ range of efficiency.

Test Setup The total quantity of water required to conduct the test is estimated and charged to the tank intended for the base suspension. A quantity of AC dust equal to 1 g/gal. is added and kept in suspension by agitation. Water is almost universally used as the base test dispersant, even though the end-use liquid will be higher in viscosity than water. Oil is frequently used in testing of filters for automotive applications.

A turbidity reading is taken on the base suspension and used as the reference reading for comparison to later values obtained as the test proceeds.

The test filter cartridge is placed in the housing and the test may commence.

TABLE 3-19. Test Dust for Evaluating Liquid Filtering Media.

Range of particle size (μ)	Fine AC dust, wt%	Coarse AC dust, wt%
80 to 200	0	9 ± 3
40 to 80	13	30 ± 3
20 to 40	18	23 ± 3
10 to 20	20	14 ± 3
5 to 10	19	12 ± 3
0 to 5	30	12 ± 2

*Supplied by General Motors AC Division.

Test Procedure   The base suspension is pumped through the filter at a constant, automatically monitored rate of 3 gal./min. Readings of turbidity of the effluent, downstream filtered water are taken every 3 to 5 min. The Hach surface-scatter type turbidimeter functions by permitting a light source shining at a low angle onto the surface to be reflected from dirt particles at or near the surface of the effluent water. A nonturbulent sump portion of the effluent tank is provided for this purpose. Light reflected by the particles is sensed by a photoelectric cell pointed directly downward at the surface. Only light reflected from the suspended particles will be sensed. The clearer the effluent water, the less light will be reflected, and the lower will be the reading, indicating a higher quantity of the test dust absorbed by the filter. Only one pass is made of the base water suspension through the test filter cartridge.

As the test proceeds, the filter naturally absorbs or removes the dust from the suspension. A fine filter will remove the dirt more rapidly than a coarse-fibered filter, hence the duration of a test evaluating fine filter cartridges is shorter.

As the filter becomes loaded, the difference in pressure (also termed pressure drop or ΔP) increases. The ΔP is continually monitored and readings taken to correlate with the turbidity values. The test for any individual filter cartridge is concluded when ΔP reaches 30 psi. The test flow rate of 3 gal./min is automatically maintained during the test even though resistance to flow increases. The ΔP of 30 psi is considered to be the practical end of usable life for the filter cartridge.

Efficiency is calculated according to

$$\%E = \frac{T_B - T_E}{T_B} \times 100,$$

where

E = efficiency;
T_B = original turbidity value of the base suspension, 1 g/gal. of test dust;
T_E = turbidity of effluent dust suspension.

The test may last 15 to 20 min for fine-fiber cartridges, and up to 90 min for the coarser-fibered variety. Figure 3-45 shows percent efficiency and ΔP plotted against time for representative fine and

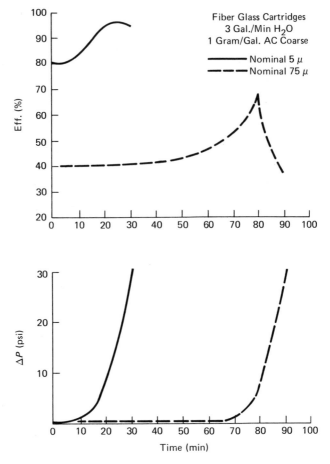

Fig. 3-45. Plot of percent efficiency and ΔP versus time of testing for fine (5 μ) and coarse (75 μ) fiber glass filter tubes. (*Courtesy Johns-Manville Sales Corp.*)

coarse filters. Several different filter cartridge units are tested, and an average value determined for purposes of reporting, or representation of performance of any specific filter type or class.

In Fig. 3-46 are presented three bar graphs showing efficiency, flow rate, and dust-holding capacity for fiber glass filter cartridges versus those made with competitive materials (wound or molded synthetic fibers).

As stated, filters are also generally rated according to their efficiency value. A fine-fibered filter provides 99% filtration efficiency

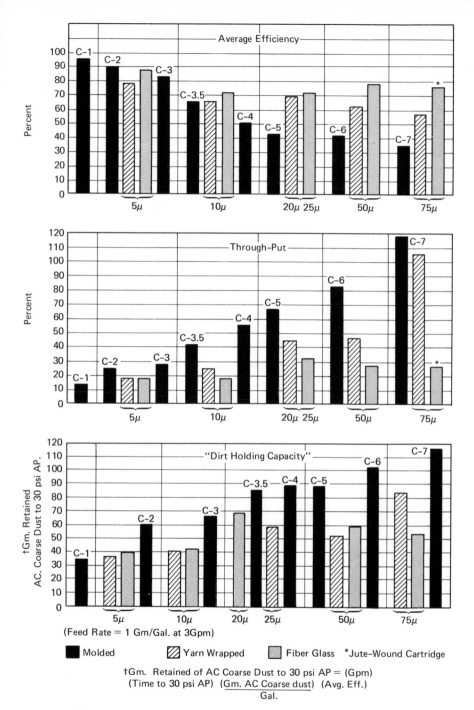

Fig. 3-46. Bar graphs showing efficiency, flow rate, and dust-holding capacity for fiber glass filter tube cartridges in comparison with competitive wound or molded (synthetic fiber) filter cartridges. (*Courtesy Johns-Manville Sales Corp.*)

of the AC test dust, while that for the coarse-fibered C-7 filter is in the range from 35 to 40%.

Applications and Performance

Because of the variety of interesting applications for fiber glass in liquid filtration, presentation is made of several exemplary areas of usage, generally moving from coarse to fine types of filtration.

1. *Paints, Varnishes, and Solvents.* In processing these commodities, it is necessary to remove dirt from the raw materials entering the reactors. These foreign particles include agglomerates, gelled and/or partially polymerized resin particles, and others. Filter tubes fabricated using D- and E-diameter or coarser glass fibers are used depending upon the manufacturer-customer agreed-upon requirements or specifications. The filters used may go up to 40 in. long, operate at 50 psi pressure, and may have from 1 to 50 tubes in series depending upon throughput volume required by the operation. (See Figs. 3-47 and 3-48).

2. *Photography Processing Plants and Laboratories.* Function of the fiber glass filters in this application is to maintain clarity of

Fig. 3-47. An example of the use of long fiber glass filter tubes in large-scale throughput processing. (*Courtesy Johns-Manville Sales Corp.*)

Fig. 3-48A. Large-scale equipment in which fiber glass filter tubes are employed for the filtration of paint. A: shows operator inserting triple-length 25 μ filter tubes into a six-element filter housing. (*Courtesy Atech Chemical Coatings Division, Wolverine Aluminum Co., and Johns-Manville Sales Corp.*)

Fig. 3-48B. Shows operator filling drums while paint flows through the filters. The smaller filter tube housing on the left side of the equipment stand is intended for the 1 μ filter cartridges for absolute liquid filtration. (*Courtesy Atech Chemical Coatings Division, Wolverine Aluminum Co., and Johns-Manville Sales Corp.*)

solutions used, remove particulates, and extend usable filter life. Materials requiring filtration in a large-scale lab include water, bleach solutions, developers, prehardeners, stabilizing (fixing) solutions, corrosive acids, and others.

Filtration down to 25 μ of contaminant particle size is required. Usually, filters are placed at individual feeder or machine locations and must filter in recycling a nominal liquid quantity of 80 gal./min. In one large central photo processing house, a total 3,200,000 gal. per year of filtered water is required.

Use of fiber glass filters in systems recirculating developer solution was found to extend the solution's usable life by 50%.

3. *Underground waterflooding.* Filtered water is pumped into dry or depleted wells, or into hard shale beds to recover or push out oil or gas still existing underground. Medium filtration (ca. 0.20 μ) is required, and differential pressures approach 75 to 100 psi.

4. *EDM (Electrical Discharge Machining).* In this relatively new, rapid, time- and labor-saving process, metal may be machined to contours too difficult for normal machine tools to handle. Savings are as much as 50 hr on one part. Oil is used as the coolant and also acts as the waste particle remover by being continuously recirculated during the "burning" process.

Fiber glass filter tube cartridges[38] removing particles 5 to 25 μ in size spell the real difference between success and failure by removing the rapidly sloughed metal particles, and preventing the sludge buildup experienced with other filter media. Glass filters last four times longer than the short 1-week life of pleated cellulosic filters. The volume of oil handled is not critical, since excessive amounts are not required in the EDM process.

5. *Filtration of Hydraulic Oil.* Removal from the hydraulic oil of all foreign matter such as metal chips, plastic fragments, etc., in any molding or pressing operation is necessary to protect costly hydraulic equipment from scoring, leakage, and other deleterious factors. Particle removal down to 10 μ is handily accomplished using appropriate fiber glass filter tubes. Other more critical apparatus may require protection down to 3 μ. Portable filtration units are available which are flow-rated up to 165 gal./min.

6. *Filtration of Swimming Pool Water.* Diatomite®, diatomaceous earth, and fiber glass cartridge-type filter tubes control 100% of the market for swimming pool water filtration, with the latter being

the newcomer. The water contains coarse dust, leaves, moths, etc., that fall in and are removed using coarse filters. The cartridges usually are triple length (approximately 30 in.), have no grooves on the outside periphery, and are covered with a porous polypropylene (not fabric) stockinette. Fine particles removed delineate to the range of haziness (5 to 10 μ). A large size public pool (40 $\times$ 100 ft) will require 50 such tubular elements, filtering up to 2500 gal./min.

The average life is 59 days per element (all year around pool) during which 0.5 million gallons have been pumped through each unit. In one comparison test on this pool unit, the 50 fiber glass filter tubes so used outperformed 150 pleated cellulosic (paper) filter tubes, and drastically reduced handling time for cleanout and replacement.

7. *Absolute Liquid Filtration.* The almost perfect 99.99% efficiency of filtration is required for processing the following: (1) removal of particles down to 1 μ from deionized water for ampule use, vial washing, and pharmaceutical use; (2) to 0.45 μ at 50 gal./min for fluorinated solvents for a variety of uses; (3) for cosmetic-base oils—down to 0.8 μ size removed using a flow rate of 75 gal./min; (4) pharmaceutical-base water—0.65 μ at 40 gal./min; (5) special paints—to 1.0 μ using 75 gal./min (see Fig. 3-48); (6) water and solvent rinses for microelectric rinses requiring 99.99% removal of particles down to 0.30 μ at 40 gal./min.

All above feats have been accomplished using fiber glass Micro-Fibers® 0.30 to 20 μ fiber diameter, codes 102 to 112.

Additionally, fiber glass is used as prefilter for micropore membrane filter elements made from mixed esters of cellulose acetate and cellulose diacetate, yielding narrow pore sizes ranging in small groupings between 0.22 and 5 μ.[39]

8. *Filtration of Jet Fuel and the Like.* As if it were not sufficient to take out of a given substance undesirable particles equivalent to the "smallest flea on the smallest flea," one would be required to perform an analysis of interatomic distances, radii, and molecular sizes to explain the capacity of fiber glass to separate and remove one liquid from another.

Water, a foreign substance in jet fuel, is undesirable, of course, and may be removed by passing the fuel through a filter made using mixtures of B-, C-, and AA-fiber glass, also with scrim fabric included for integrity. The water exists in the fuel in an extremely finely divided state.   The phenomenon accountable for its removal by filtration is termed "coalescence."   As the fuel passes through the fiber glass filter cartridge, the water collects on the fiber, slowly filling pores as droplets enlarge, then gradually collecting on the surface and moving downward by gravity and out a trap while the jet fuel rises and passes through the housing exit provided.   The following physicochemical properties of the filter media are responsible for coalescence; Fiber glass by itself is hydrophylic (high capillary attraction).   What is required is a balance in the media in which it is rendered neither strongly hydrophobic nor hydrophylic. This is accomplished using special binders and mixed fiber sizes.   In filtration, water is taken from the fuel and held on the fiber surfaces until the ΔP reaches a value greater than that of the surface tension between water and fiber.   At this point the water is pushed through, collects in droplets, and removes itself as described.   Fortuitously, this constitutes a reversible reaction, and coalescing filters experience well-extended life cycles.

FIBER GLASS MAT AND WEB PRODUCTS

Introduction

The many-sided versatility of fiber glass is further illustrated by its eminent performance in a series of applications essentially involving or starting with flat mats or webs.   The forms of fiber glass used are quite varied and result from several fabricating methods, mostly blown fiber, but including some continuous-filament or textile products (see Chapters 4 and 5).

The applications themselves include the following: papers; shingles and other roofing products; industrial mats for pipeline covering, etc; decorative and functional mats; battery separators and/or retainers; surfacing and veil mats; and other relevant uses.   Each application or series will be detailed and discussed following the order outlined above.

Glass Fiber Paper

Manufacture of papers either combining glass fibers with bleached sulfite, cotton linters, hemp, pine, or other types of cellulosic pulps, or using 100% glass fibers, is well established.   When glass is added to cellulosic pulps, almost all processing and ultimate physical properties are improved. Conversely, if used 100% as the slurry or stock ingredient, resultant papers possess improved porosity and temperature resistance related directly to the chemical composition of the base or modified glass.

In Table 3-20 are recorded the results of hand-sheet tests in which 2 to 20% of various glass fiber (microfiber and 9 μ G-fiber, continuous-filament chopped strand) was added to bleached sulfite pulp. Note that improvements resulted in all wet-process parameters, and in all dry paper characteristics excepting dry tensile strength and fold endurance.[40]

In practice, additions of glass fibers to cellulosic paper stocks generally lend improvements based on the parameters discussed

TABLE 3-20.   Effect of Glass Fiber Additions to Bleached Sulfite Pulp (Cellulosic Paper Stock).

Property (units)	Effect of 2–20% addition	Detailed resumé
Wet tensile strength (oz/in.)	Improved	Code 112 best at 2% (12 oz/in.) Code 108 best at 20% (12 oz/in.) (cellulosic original = 5 oz/in.)
Wet tear factor (oz/in.)	Improved	9 μ fiber best—from 80 @ 0% glass to 180 @ 20%
Wet bulk (cm³/g)	Improved	9 μ best—to 24 @ 20% Code 112 = 21 @ 20% (original = 17.2)
Drainage	Improved	All glass types improved
Brightness and opacity (Tappi T-452-m58)	Improved	Code 106 = 84 @ 20% (original = 76.5)
Dimensional stability	Improved	9 μ and coarse fiber best
Dry tensile strength	No improvement	Small % additions = no decrease Large additions = no improvement
Fold endurance (MIT double)	No improvement	Original = 75 folds All glass additions reduce to 10 to 30 folds
Burst strength (Mullen)	Some improvement	Larger filament diameters show increase to 25% glass addition

above.   Specific end uses include papers for electrical insulation, permanent documents, plastic laminates, and others.   Glass fiber additions increase processing speeds, hence offset additional costs.

Complete, 100% glass fiber papers can be fabricated using the various types listed in Table 3-21.[40]   Process improvements necessary to gain increased temperature endpoints are indicated where necessary.

As stated, the prime advantages of 100% glass fiber papers are in the increased percentage of voids.   When used for filtration of either gases or liquids (see Filtration) there is no abrupt, limiting deleterious rise in ΔP for approximately 75% of the total cycle time (filter life). Although filtration accounts for the major use of 100% glass fiber papers, many specialty uses are developing such as for high temperature gasketing, electrical insulation, sound absorption, and nuclear thermal control.

TABLE 3-21.   Type, Fiber Diameter, and Thermal Resistance of Various Fiber Glass Types Used in Paper Manufacture.

Material	Composition or other identification	Average fiber diam (μ)	Temperature resistance (°F)
Micro-fibers®*	475 glass*	0.05 to 3.8	800
Micro-fibers®*	E glass	0.5 to 1.6	1000
C to D fibers	Insulation glass composition	3.8 to 6.4	800
Micro-quartz®*	Leached to 98.5% SiO_2	0.5 to 3.8	2850
Chopped strand	E glass, (continuous filament)	10 to 14	1000
SG (spun glass)* fiber	800 glass	3.5	800
Ceramic refractory fiber (cera-fiber)*	Alumina-silica composition	3 to 4	2300
Fiber chrome* refractory fiber	Alumina-silica chrome-oxide fiber	3 to 4	2700
SA ceramic fibers	–	0.5 to 2.6	2000

*Designations of Johns-Manville Sales Corp. products.

Shingles and Roofing Mats

Shingles

Although glass fiber products (mats and boards) have been used in roofing products in old or new construction for a considerable length of time, their entry into shingle fabrication is of fairly recent origin. Whereas mats of organic material, or of asbestos were previously saturated with asphalt in forming shingles, glass fiber types have proven to be more effective by virtue of the following: (1) more rapid rate of saturation by the asphalt component, permitting faster production line speeds; (2) more thorough saturation than organic felts or other competitive shingle mats, eliminating tendencies for post-manufacture moisture absorption, shingle distortion, blistering,

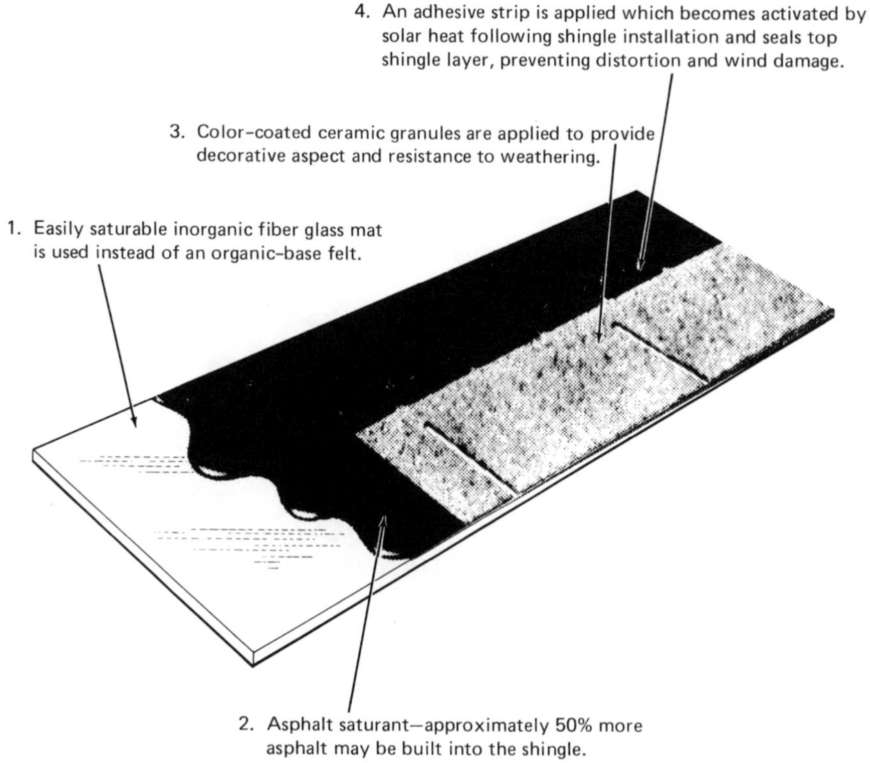

4. An adhesive strip is applied which becomes activated by solar heat following shingle installation and seals top shingle layer, preventing distortion and wind damage.

3. Color–coated ceramic granules are applied to provide decorative aspect and resistance to weathering.

1. Easily saturable inorganic fiber glass mat is used instead of an organic–base felt.

2. Asphalt saturant—approximately 50% more asphalt may be built into the shingle.

Fig. 3-49. Composite view showing steps and materials required in construction of a modern fiber glass shingle. Elements are described within the body of the illustration. (*Courtesy Johns-Manville Sales Corp.*)

and other undesirables which could cause premature failure. The incorporation of fiber glass also yields a shingle capable of a UL Class A fire rating versus a lower Class C rating for shingles made using organic felts.

Glass mats suitable for incorporation into shingles are selected according to weight per square (100 sq ft). This may be any ultimate value up to 3 lb/square depending on final finished weight of the shingle, the highest for which usually approaches 250 lb/square. Edge reinforcing, high tensile strength, and flatness (freedom from wrinkles) are other superior attributes of the base fiber glass shingle mat.

Figure 3-49 illustrates the relationship of the four main elements in modern shingle fabrication, i.e., glass fiber mat, saturating asphalt, colored ceramic granules, and adhesive sealing strip.

The advantages of using fiber glass saturating mats in shingle construction are clearly evident in Fig. 3-50. The X ray or microradiographs in cross section of fiber glass and organic felt shingles show the distinct differences in degree and completeness of saturation.

Originally, some difficulty was encountered in cold-weather cracking of fiber glass shingles. This was corrected and eliminated by over-bulking the fiber glass saturating mats by a factor of 40 to 50% while maintaining finished shingle equivalent weight.

Fig. 3-50A.   Cross-sectional X-ray photomicroradiograph of an asphalt-saturated fiber glass shingle. Irregular contour of the ceramic granules is obviously visible at the top of the illustration. The totally dark area throughout the body of the photo points up the 50% greater degree of saturation of the fiber glass mat by the asphalt impregnant. This eliminates water absorption, blistering, distortion, premature failure, and also permits the fiber glass shingle to carry a UL Class A fire rating. (*Courtesy Johns-Manville Sales Corp.*)

Fig. 3-50B. X-ray microradiograph of a shingle made using an organic felt. The light or mottled area in the body of the shingle shows evidence of incomplete saturation, typical of all organic-based shingles. In service, this may permit moisture absorption, contributing to distortion, blistering, and premature failure. Also, the incompletely saturated organic felt shingles carry only a UL Class C fire rating. (*Courtesy Johns-Manville Sales Corp.*)

Built-Up Roofing

On flat or slightly inclined roofs of commercial or industrial buildings, fiber glass has contributed substanially toward solving the long-term leakage and deterioration problems induced by temperature differences and moisture. Roof temperature differentials of 185°F (-20°F winter to +165°F induced by summer sun) cause linear expansion or contraction of as much as $\frac{19}{64}$ in. per 10 ft span if the supportive roof elements happen to be aluminum.[41]

Main construction materials for flat built-up roofs may be wood, expanded steel or other metal, lightweight concrete, or poured gypsum, usually 2 in. in thickness. In the technology developed for better roof construction, fiber glass board or batting is used for thermal insulation both below and above the main structural roof element (see Fig. 3-51).

Following installation of the structural members and the proper top board-type insulation, thin fiber glass mats are applied and saturated with hot-steep asphalt. The makeup of the builtup roofing layer, usually $\frac{3}{8}$ in. in finished thickness, is two or three layers of a fiber glass "ply sheet" and one layer of a "cap sheet." The ply sheets are usually fully saturated, and weigh 1 to 2 lb/square before saturating and about 15 to 20 lb/square after saturating. The cap sheet also

Fig. 3-51. Placement of board-type fiber glass insulation preparatory to casting poured-in-place roof deck. (*Courtesy Owens-Corning Fiberglas Corp.*)

is preimpregnated and may have a finish of fine ceramic grains similar to that applied to shingles. In Fig. 3-52 is illustrated application of layers of saturated fiber glass mats in a built-up roofing fabrication.

Many accessories and variations are possible. Narrow tapes of fiber glass are made for impregnation around flashing, etc., at the edges of roofs. The insulation products are sold with vapor barriers affixed for greater facility in construction, or if not, the vapor barriers are included below and above the thermal insulation layers as extra applications.

Industrial Bonded Mats

Pipeline Protection

Corrosive electrochemical attack of underground metal surfaces due to action of stray cathodic currents is critical in any location, but usually worse in densely populated areas. Submerged steel tanks for gasoline storage have had leakage problems delayed by coating them with a fiber glass reinforced plastic layer for protection. More

Fig. 3-52. Application of rolls of fiber glass ply sheets in built-up roofing. (*Courtesy Schuller GmbH, Wertheim, Germany, Subsidiary Johns-Manville Corp.*)

recently, metal underground storage tanks have been largely super-seded by all-RP/C tanks (see Reinforced Plastics).

With the increased usage of underground steel piping to transport oil, gas, and some chemicals, the corrosion and concomitant leakage problems were even more critical due to the complexities of deter-mining the actual point of material loss over the path of a long, multibranched distribution system.

Again, a form of fiber glass was pressed into service and, used in conjunction with fiber glass reinforced asbestos felt, provided the basis for a superior pipe protection system.

The fiber glass "bonded" mat contains phenol-formaldehyde or polystyrene binder, and is usually reinforced by longitudinal con-tinuous-filament fiber glass strands which are laminated in. Mats are usually 1.05 lb/square (100 sq ft), 0.020 in. thick, provide a tensile strength (dry) of 15.0 lb/in. (lengthwise direction), and the rein-forcing strands are on $\frac{1}{4}$ in. centers. The base fiber glass composition employed is a medium chemically durable lime-borosilicate type.

Fig. 3-53A. Pipe-wrapping machine in operation. The narrow-width rolls of glass mat and asbestos overwrap are applied in a helical wind from appropriately stationed spindles. (*Courtesy Johns-Manville Sales Corp.*)

During application, the pipe is saturated with a hot tar or asphalt "enamel" which sets upon cooling after the glass and outer protection mats have been wrapped on. Application may be made either in a storage or distributing yard, or in the field. Figure 3-53 illustrates machinery designed for field pipe coating in action.

Specialty accessory products are also supplied such as tar-impregnated fiber glass tape for wrapping pipe joints (to be heated as applied) and combined glass-asbestos-asphalt mats for heavy-duty overwraps and rock shields.

Roadbed Protection

Overseas moreso than in the U.S., bonded fiber glass mats have been used to refurbish asphaltic and other roadbeds and protect them from further washouts and freezeouts during periods of inclement

Fig. 3-53B. Progression of the operation of protective pipe covering in a field application. (*Courtesy Johns-Manville Sales Corp.*)

weather.   Figure 3-54 shows an operation being carried out in Germany.

Drain-Tile Protection

In the flat, fertile farmlands of the American midwest, especially in areas neighboring the Great Lakes, it is necessary to establish an elaborate drainage system using clay tile and run-off ditches. If sown fields were not tiled, flooding induced by heavy rains would persist, and cause needless crop damage.

Part of an area farmer's economic success is measured by the infrequency in which he is forced to lay new tile, because those last installed had become clogged due to his good farm soil being washed in.

Very recently (ca. 25 years) a combination farmer and fiber glass plant worker conceived the idea that clogging might be prevented by placement of a single layer of fiber glass industrial bonded mat over freshly laid drain tile before replacing dirt in the excavation ditch.

Fig. 3-54. Bonding of a fiber glass mat to a roadway surface using hot-steep asphalt for protection against wear, expansion, contraction, washouts, freezeouts, etc. (*Courtesy Schuller GmbH, Wertheim, Germany, Subsidiary Johns-Manville Sales Corp.*)

The results were positive, and opened the way for "mucho" fiber glass industrial mat sales, plus "poco" frustrated area farmers.

Backing for Floor Tile, Carpeting, and Wall Covering

Jute ground fabric and associated natural fibers are widely used in not only primary facing but in secondary backing for carpeting and individual software floor tile. Regardless of pretreatment used, these organic materials possess excessively high moisture sensitivity and resultant shrinkage in use.[42]

A method has been devised for applying a PVC coating to the types of industrial fiber glass mats currently under discussion to form backings possessing great dimensional stability. Such a coating system is illustrated in Fig. 3-55 (see Fiber Glass Textile Fabrics).

Glass mats used range from 1.0 to 1.6 lb/square, 0.018 to 0.028 in. thick, with machine-direction tensile strengths up to 25.3 lb/in. The

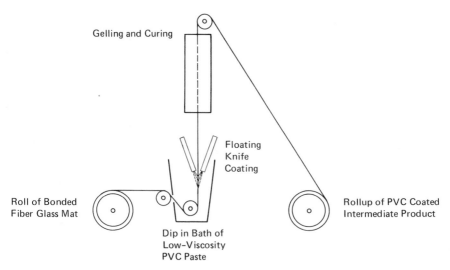

Gelling and Curing

Floating
Knife
Coating

Roll of Bonded
Fiber Glass Mat

Rollup of PVC Coated
Intermediate Product

Dip in Bath of
Low–Viscosity
PVC Paste

Fig. 3-55. Schematic drawing representing process for application of PVC coating to fiber glass mats for tile and carpet backing or wall covering. (*Courtesy Schuller GmbH, Wertheim, Germany, Subsidiary Johns-Manville Sales Corp.*)

original binder is a modified urea-formaldehyde type, providing good compatibility with the PVC plastic, and offering high porosity for rapid impregnation. Other components necessary to form the wall covering, tile, or carpet backing may either be included with the coating process, or post-applied.

Battery Retainer Mats

Separator Sheets for Small Batteries

Electrical lead-acid type storage batteries, principally for automotive use, consist of anodic (+) and cathodic (−) poles in the form of closely spaced plates submerged in a H_2SO_4-H_2O mix inside a suitable corrosion-resistant closed container. While in action (discharging), the anode is chemically reacted from lead peroxide to lead sulfate and the cathode from pure lead to lead sulfate. The electrical output is approximately 2 V per cell.

The reactions described reverse themselves when the battery is being charged. Although the battery's function is supported by the car alternator (generator) during running time, the chief mode of the battery during its life is in discharge. Since lead sulfate ($PbSO_4$)

has a higher molecular weight than either lead peroxide (PbO_2) or lead (Pb) it has also a higher volume percent and hence occupies more space.

Unless retained, the sloughing off and building up of the lead sulfate particles will cause a short circuit to occur between the battery plates. Many materials including organic fibers have been used in the past as separator mats between the cells to retain the sloughed $PbSO_4$. However, none have performed so capably as fiber glass, which has dominated this market almost since its commercial availability in 1937.

The three most salient advantages of glass fibers for use in the separator mats for smaller batteries are: (1) chemical durability provided by use of the C-glass composition[43] to retain structural mat integrity in the presence of 1.3 sp. gr. H_2SO_4; (2) capability of being fabricated into a proper shape, made rigid by clay-base binders; and (3) proper cell or pore size to permit transfer of the electrolyte, yet retain the sloughed lead sulfate.

In Fig. 3-56 is exhibited an array of fabricated fiber glass separator mats for use in automotive storage batteries.

Fig. 3-56. Battery separator sheets for use in small electrical storage batteries of the automotive type. (*Courtesy Johns-Manville Sales Corp.*)

Laminated Battery Separator Mats for Larger Batteries

Whereas a single 12 V output battery will handle the running requirements of a standard automotive vehicle, and three to six or more similar-type batteries will power small electric cars, much heavier duty lead-acid batteries are required as prime movers in industrial forklift trucks and similar equipment. In industry, the electrically driven service vehicles are far more desirable than their counterparts driven with combustible fuels, because of toxic, polluting exhaust gases, noise, etc., generated by the latter.

However, the electric batteries are much larger in size than the automotive type, and quite heavy. This is not entirely a disadvantage because greater counterweighting of applied loads is frequently needed for this type of vehicle.

In comparison to battery plates roughly 6 X 6 in. used in the automotive batteries, plates required for the lift truck batteries average 6 X 14 in. Sizes even up to 24 in. long are contemplated for some additional uses described later in this section.

The fiber glass industry has kept pace with technological progress, and worked together with the battery manufacturers to meet the additional demands of these larger battery units. Since the plates were larger and thicker, and battery life up to 6 years was desired, a tighter-fitting fiber glass battery-plate wrapper or retainer mat was required.

This problem was solved over the years by utilizing and combining two forms of fiber glass into a laminated wrapper element: (1) a fiber glass industrial-type mat bonded with either acrylic or polystyrene resin and weighing from 0.8 to 3.5 lb/square (10 to 60 mil thick), depending upon battery type; and (2) a staple-fiber strand, termed "sliver," also of C-glass, g-filament (9 μ) diameter and of approximately 620 yd/lb denier, and without twist.[44]

The sliver strands are flat and, working directly from creeled textile packages, are drawn onto and adhered to the mat in a moving or throughput operation at a spacing of 4 to 5 strands/in. The strands are approximately $\frac{1}{4}$ in. wide so that almost complete double glass coverage results. However, the resultant laminated mat has the required high permeability. The sliver strands are laid on parallel in one direction only, and do not cross.

The combination of these two fibrous glass types has, more than

(A)

Fig. 3-57. Details of construction of a lead–acid electrical truck battery showing use of fiber glass separator mat. Inset view, Part A, shows detail of a laminated mat with sliver adhered to the mat base. Part B shows a cutaway view of a
unit cell with individual components referenced and annotated as follows.

1. Vent cap for watering and inspection.

2. Lead insert bushing in battery cell cover to prevent leakage and grounding
of output voltage.

3. Cover panel to guard and protect plates and separator mat from being
damaged when testing specific gravity of the acid-water mix using a hydrometer.

4. Contactor posts connected to positive and negative battery plates.

5. Negative plate.

6. Positive plate (upper portion) showing grid construction.

7. Positive plate (lower portion). Active lead-base material is molded into
plate grid (construction of negative and positive plates is essentially the same
grid and active filler combination. Composition of active filler differs for each
plate).

8. Fiber glass battery retainer (separator) mat consisting of staple fiber (sliver)
facing the active plate material and adhered to the mat. The retainer mat is
wrapped vertically around the plate. This retainer permits free passage of the
electrolyte, but prevents transfer of particles of active material which might
become dislodged, thereby precluding shorting-out and providing battery life
for periods of at least 6 years.

9. An additional horizontally wrapped glass mat with a comparatively looser
construction and a chemically insoluble binder. The purpose of this component
is to break up gas bubbles formed, and also increase positive plate insulation.

10. Perforated, tight-fitting polyethylene (or equivalent) plastic sheath. This
component encases all positive plates and glass mats to establish maximum performance and extended battery life.

11. Porous, ridged, and channeled inorganic membranes which further separate and insulate the plates, yet permit free flow of electrolyte through the cell.

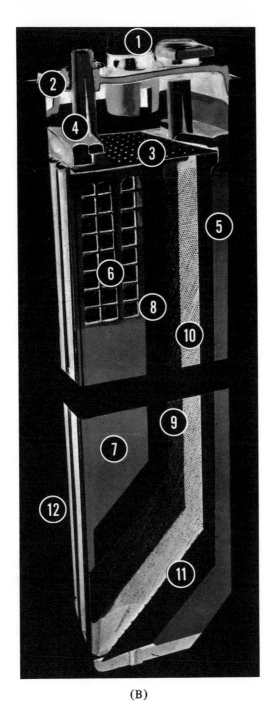

(B)

12. Cell container or jar, molded of high-impact chemically resistant material so as to be tight and leakproof. (*Courtesy East Penn Mfg. Co. and Sun Chemical Corp., Facile Division.*)

any other factor, accounted for extension of battery life to periods up to 6 years.

The retainer mats are mounted onto the battery plates in a vertical aspect with the sliver side toward the active Pb or PbO_2 plates. The retainer mat is not loose but tightly fitted onto the plate and encased by a perforated plastic screen, usually polypropylene. Figure 3-57 shows a sample of battery mat plus a cutaway view of an actual cell of a typical large-size, high-performance battery.

Batteries in this class contain from 12 to 24 individual cells. Ratings range from 12 to 72 V and 150 to 2175 ampere-hours. In addition to usage in industrial fork lift and associated gear, this same type of battery is also employed as the main power source in telephone lines or stations, and for submarines.

One unusual contemplated use that is well on the way toward reality is that of a load leveler for a large electric power utility. A group of heavy-duty lead-acid batteries will be connected in series and cover an area approximately the size of a football field. They will be connected to and charged by the mechanical power generating units. When peak loads occur, the batteries will be discharging, and conversely, will be charging during nonpeak loads. Hence, the

TABLE 3-22. Battery Types Competitive to Lead-Acid and Their Projected Dates of Commercialization.

	ENERGY SOURCE COMPARISON (Projected Characteristics and Availability)				
Battery	WH/lb	W/lb	$/kwh	Cycle life	Projected availability
Group I					
Lead-Acid	16–18	25	20–50	300–2000	Now
Nickel-Iron	20–30	60	20–30	300–400	1978
Nickel-Zinc	30–40	75	20–25	250–350	1978
Group II					
Zinc-Chlorine (Hydrate)	60	60	10–20	500*	1978–79
Lithium-Sulfur	60–80	100	15–20	1000*	1980–85
Sodium-Sulfur	80–100	100	15–20	1000*	1980–85

*Nominal life based on current experimental models.
(Source: *Independent Battery Manufacturers Assn., Inc.*, 100 Larchwood Drive, Largo, Fla. 33540.)

utility power station will be able to operate continuously at a nominal rate, and wide load variations required by intense peaking out or low slack times will be virtually eliminated or at least greatly ameliorated. The plate sizes required by these batteries are up to 24 in. long, hence it is easy to visualize the additional amounts of fiber glass required for retainer mats.

One associated point of interest concerns the threat to the currently strong commercial position of lead-acid batteries by competitive types. In Table 3-22 are listed battery types under development, and yearly dates estimated for their commercialization. The consensus is that fiber glass-assisted lead-acid batteries as a viable power source will enjoy a considerably long and healthy future.

Veil and Surfacing Mats

Glass fiber mat products made by the processes schematically depicted in Figs. 2-9A, 9B, and 2-10 comprise thin, lightweight materials which function in reinforced plastics/composites (RP/C) to provide both surfacing and reinforcement for molded plastic structures (see Chapter 5).

On one occasion in the early period of reinforced plastics, a small manufacturer received a purchase order from a chemical company to fabricate several reactor cylinders. These were to be approximately 2 ft diameter × 15 ft long. Standard general-purpose polyester and fiber glass reinforcement were specified for the main tube structure. However, due to the chemical nature of the material to be reacted, an epoxy resin coating on the inside surface of the cylindrical reactors was also specified.

The fabricator, knowing that epoxy resin was easily handleable and had excellent adhesion, applied it by brushing to the inside of the tubes after they had been molded.

As you can well predict, if you have any familiarity with these materials, after only a few hours in service, the inner epoxy coating separated itself from the body of the laminate and floated to the top of the liquid, rendering the reactors useless. A thin layer of fiber glass veil mat molded into the epoxy resin layer would have precluded this debacle.

This type of experience was probably repeated on many occasions, and led to the specification and use of veil and surfacing mats in

(a)

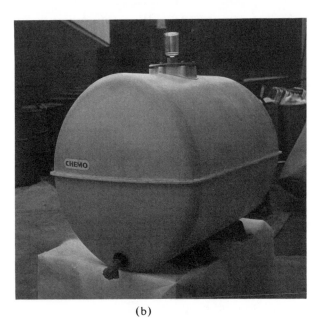

(b)

Fig. 3-58. Automotive (Part A) and corrosion-resistant (Part B) reinforced plastic structures in which veil and surfacing mats have been used to improve appearance and performance. (*Courtesy Schuller GmbH, Wertheim, Germany, Subsidiary Johns-Manville Sales Corp.*)

fabrication of all corrosion-resistant RP/C structures. Veil mat is saturated with the required corrosion-resistant resin as the preliminary step, to be followed by the main structural layup. The veil mat provides only about 10% glass content for that particular resinous layer, but assures stability and bonding to the rest of the layup.

Veil and surfacing mats are also widely used in RP/C molding processes to assist in elimination of voids and wrinkles, and to eliminate patterning caused by the reinforcing fibers.

Veil mats contain little or no binder and have high drapability. They are somewhat difficult to handle, but provide excellent results. Surfacing mats contain a finite percentage of a polyester-compatible binder, and are more resistant than standard reinforcing mats to washing during lay-up or molding.

Both veil and surfacing mats are made in thicknesses from 0.010 to 0.030 in. (approximately $\frac{1}{8}$ to $\frac{1}{2}$ oz/sq ft).

When permitted to accumulate in weights of approximately 1 to 2 oz/sq ft, these mats, particularly the mechanically attenuated type (Fig. 2-10), are used as reinforcing fibers in molding noncritical RP/C items such as TV trays, refrigerator drip pans, and the like. Drapability is excellent and interesting decorative effects result from the characteristic diamond-shaped fiber pattern.

Figure 3-58 illustrates use of veil and surfacing mats in fabrication of automotive and corrosion-resistant RP/C components.

Chapter 4

Continuous-Filament Fiber Forming Methods

INTRODUCTION

Historically, the continuous-filament type fiber glass antedates the wool variety by quite a few centuries. Syrian and like period glassmakers pulled fine threads onto tooled or off-hand glass objects for purposes of decoration. Venetian glass artisans incorporated different colored glass strands (mostly white) in crossed or weblike configurations into the body of their fine, thin-walled objects (see Fig. 4-1). Both French and German craftsmen in the mid-18th century learned to produce separate fibers of glass, the German process drawing monofilaments to the side from a hot melt.

Numerous laboratory curiosities simulating fiber-drawing equipment were as common as the earlier Prince Rupert drop or Bologna phial.* A resistance-heated platinum strip was arranged with a hole in the center, slightly bent into a boat shape to hold molten glass, and a small hole in the center through which the melted glass would exude. When heated, the glass could be pulled away as a tiny filament by action of a high-speed winding drum.

Just before the turn of the century in 1893, the Columbian Exposition was held in Chicago in which the most effective glass entrepreneurs in U.S. history, Messrs. Edward Drummond Libbey and Michael J. Owens, exhibited many utilitarian glass items including a fiber glass dress. See Fig. 4-2.

As stated, commercially important continuous-filament fiber glass products and technology resulted from the joint Owens-Illinois and Corning Glass works research which culminated in formation of a manufacturing facility in 1937. The entire field has expanded at an

---

*Small glass novelties, highly stressed or tempered, which would explode catastrophically when the tensile stress area was penetrated.

Fig. 4-1. A late 16th century form of Venetian "Cristallo" art glass termed "latticinio" in which fine threads of opaque, white "lattimo" (milk glass) were deftly and skillfully blended into the body of off-hand glass creations to produce spectacular crisscross and other designs such as this one. Later this technique was further refined by coincorporating and attenuating minute bubbles in a clear matrix of the white lattice. The result was the well-known "lace glass." (*Courtesy Toledo (Ohio) Museum of Art, "Vase mounted as Ewer," Exhibit No. 60.36, Venice, about 1600.*)

enviable rate between 15 and 25% per year almost every year since inception.

The development of continuous-filament forming methods is documented in this section. Other related methods of producing fiber glass textiles are included. Major end-use applications are described in Chapter 5.

MARBLE MELT PROCESS

The very first fiber glass bushings contained 51 holes or tips (50 for use and 1 spare tip). Marbles from a separate melting operation were fed to a crucible or fabricated platinum bushing, which was oriented vertically and about 7 in. high. The E-glass borosilicate

Fig. 4-2. Photograph of dress made from rod-drawn glass fibers and exhibited at the Columbian Exposition, Chicago, in 1893. (*Courtesy Toledo (Ohio) Museum of Art.*)

composition, found necessary to improve over the properties of A-soda-lime glass, dates from the 1930–35 period. See Table 5-1.

Bushings containing 102 and 204 holes were developed in rapid succession, and standards for weight-length relationship (yards per pound, strand count, or glass "cut") were established (see Table 3-2 and ASTM Specification D578-61).

The base yardage, designated by letter and coordinated with filament diameter range, was originally related to yards per pound for a 200 filament strand. The Tex designation (inverse to yards per pound) is now of international importance, and is summarized for reference in Table 5-2.

Figure 4-3 presents a schematic diagram of a marble melt process, with the glass tank marble production visibly separated from the marble melt bushing. The bushings are aligned in rows and marble distribution stations located above for gravity feed. Bushing temperatures are nominally 2300°F for remelting E-glass. Following fiberization, the glass filaments are passed over a roller or belt mechanism for organic or other sizing application. These sizing solids are present in amounts less than 2% by weight of the finished product, but dictate end use of the glass fiber product. Sizings require cure by oven drying prior to secondary processing.

The filaments are directed conelike into the vortex of a mechanical gathering device to establish the actual strand. Traversing mechanisms rapidly oscillate the strand as it is drawn into the winding drum, the latter drawing the filaments downward at approximately 2 miles/min surface speed. This high speed is necessary for attenuation to required filament diameters.

As time and technology progressed, the number of bushing holes was increased and portions of the filament bundle were subdivided or

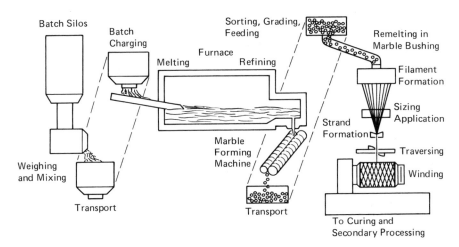

Fig. 4-3. Schematic representation of marble melt process for production of continuous-filament fiber glass.

"split" to form finer-denier yarns within the same or separate form-ing packages.

The direct-melt process (1950) was a normal consequence to technology gained in marble melt, with 400 and 800 hole bushings employed. However, one parting shot interjected by the marble process was the development of bushings with excessively high tip numbers. These were employed for production of fine-filament B (Beta)® fiber (3.2 μ nominal filament diameter). The reason was that larger bundles of filaments were required for proper denier of the fine B-yarns which would provide "soft" feel and desirable handling properties required for decorative, bedding, and clothing yarns (see Fabrics).

The technology acquired with the B-fiber development in marble bushings was of distinct assistance in development of larger bushings and creation of more versatile fiber glass product types from the direct-melt process.

DIRECT-MELT PROCESS

In Fig. 4-4 is presented a schematic diagram of the direct-melt process. The bushings are aligned on the underside of channels and forehearths directly connected with the glass-melting furnace.

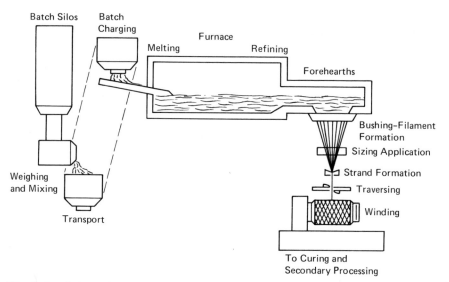

Fig. 4-4. Schematic diagram of the direct-melt process for production of continuous-filament fiber glass.

Whereas the glass depth within the furnace is 36 to 48 in., depending upon size, glass level height over the fiberizing units is only approximately 9 in.   Hence, lesser dimensional bushing heights in direct melt are required than those for marble melt.

Bushings are arranged so that the fiber-winding mechanisms exist in rows in a "forming room" with vertical fiber-drawing components similar to those for the marble process.   Automatic fiber-conveying means out of the forming room are provided.   Water, the universal solvent, coolant, and lubricant, is copiously applied at several points in the process to assist the fibers in being drawn over the mechanical components without damage.   Figure 4-5 shows an actual view of a

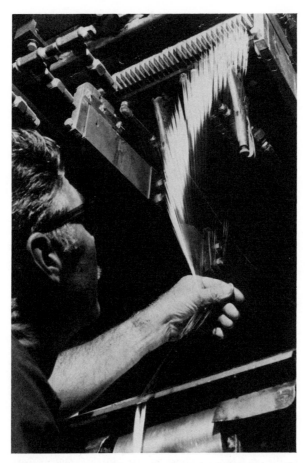

Fig. 4-5. Filaments from a direct-melt type bushing being threaded-up in preparation for high-speed drawing and fiber glass strand formation. (*Courtesy PPG Industries.*)

direct-melt type bushing being "threaded up" for drawing fibers up to speed and down to specified filament diamters.

Direct-melt type bushings up to 2400 and more tips are now extant, with larger, better performing units on the drawing boards.

Several interesting variations have been spawned as adjuncts to continuous-filament fiber glass production methods. It has always been desirable to produce larger or more usable strands direct from the fiber glass production bushings. Hence, technology for producing multifilament roving packages (see Chapter 5) and also for direct chopping into short fiber lengths from the bushing has been devel-

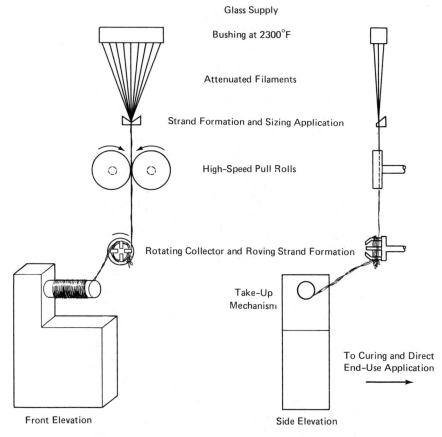

Fig. 4-6. Schematic representation of the process for producing twisted or "spun" roving direct from the bushing. (*Courtesy Owens-Corning Fiberglas Corp.*)

oped. Previously both were processes requiring interim drying and finishing steps.

Also, processes were devised for winding a single, 204 or 408 filament strand into a "spun" roving package (see Fig. 4-6) and for forming a continuous-filament mat using oscillating impact plates.

THE STRICKLAND PROCESSES

Unique attempts to greatly reduce quantities of platinum required for glass fiberization were successfully achieved by one inventor.[45] Originally a pressure system and ultimately a gravity throughput method, the second most important design parameter was the great increase in tip density, or number of holes per unit of fiber-producing area. The process is highly proprietory and probably not yet the source of a major fiber glass production facility.

FIBER PRODUCTION FROM CERAMIC CRUCIBLES

Numerous attempts have been made to produce fiber glass at lower cost from melts made in fired clay or ceramic crucibles. The glass source has been either marbles or cullet remelted by a combination of gas and electric booster melting. An inconel melting plate has been used instead of any platinum. Minimum practical filament diameter is 25.4 μ (0.001 in.), and the clay pots are fairly short-lived. One advantage is that filaments which become broken out and bead down can be refed into the main strand.

The latest attempt is well documented in one of the recent SPI RP/C Technical and Management Conference Proceedings.[46] (See Fig. 4-7.)

METAL-COATED GLASS FIBERS

Many combinations of glass or glass fibers and ceramics with metals have been proposed and developed. The main objects have been to render a normally dielectric material partially conductive for certain electrical applications. Other applications of military importance have been developed. The processes are proprietory. Further discussion of the resultant end products will be made in Chapter 5.

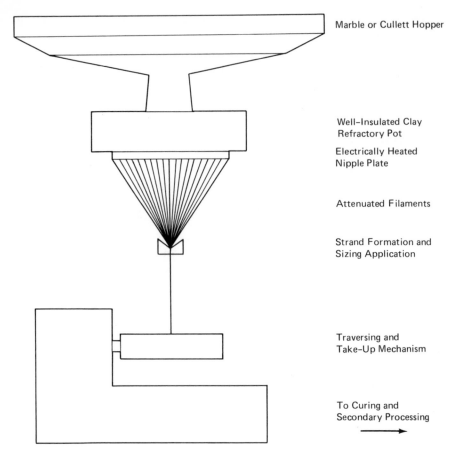

Marble or Cullett Hopper

Well–Insulated Clay
Refractory Pot

Electrically Heated
Nipple Plate

Attenuated Filaments

Strand Formation and
Sizing Application

Traversing and
Take–Up Mechanism

To Curing and
Secondary Processing

Fig. 4-7.   Schematic diagram of a process for production of continuous-filament fiber glass using ceramic (clay) type bushings with base-metal tip plate. (*Courtesy of SPI RP/C Institute*.)

STAPLE FIBER OR SLIVER

The development of sliver (at least in America) occurred during the period of early experimentation after wool products were known and prior to finalization of continuous filament.

After it had been shown that blown fibers could be produced by attenuation of fine streams of glass using an air or steam blast, and prior to the complete adaptation of platinum as a crucible material, the glass melting for fiberization was carried out in small furnaces made of clay and like refractory materials into which holes had been drilled.   Difficulties were experienced with continuing production in the clay refractory crucibles due to enlargement of the holes by action of the hot glass.   This naturally changed the blown sliver properties, and in time, even the ability to attenuate the glass. Also, the sliver product was unacceptably low in strand tensile strength. Continued action by the group looking for higher-strength fibers resulted in the final successful adaptation of platinum and ultimately the continuous-filament fiber glass which provided a more adaptable textile material.

It was discovered that if the random fiber accumulation from a single wool unit was slowly removed in a lineal path, and a slight superficial twist applied, a loosely carded glass sliver strand resulted. This could be compacted, twisted, plied, etc., to form strands for weaving.

While one faction of the groups involved in fiber glass development preferred to utilize this readily available strand woven into cloth and used for electrical laminates, the other group sought to press for development of the continuous-filament material.

The platinum technology was, of course, eventually applied to the applicable wool-producing units, and use of sliver continued for electrical and other textile uses in which high strand tensile strength was not the prime requisite.   The original production equipment for staple fiber has not changed appreciably since its inception.

Staple fiber development in Europe occurred just before or during World War II in Germany (by Schuller) when asbestos substitutes were drastically needed.   A process was developed in which fibers from a 1 m long bushing were attenuated by action of a drum revolving at high speed, with its main shaft located in the same plane as the longitudinal axis of the bushing.   The fibers wrapped around the face of the drum and were removed using a doctor blade and passed to the side through cones in which the loose-packed sliver strand was formed.

In Fig. 4-8 is presented a schematic view of one characteristic process for production of staple fiber yarn.   Fibers are rapidly attenuated downward from either a marble or batch-reduction melt, and are collected by air suction or other means along a moving belt or drum.   They are then immediately formed into a loose, bulky strand.

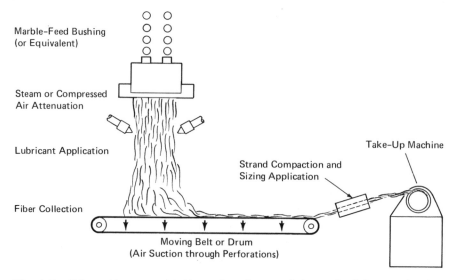

Fig. 4-8. Schematic representation of a characteristic method for production of staple fiber or sliver.

The fibers are approximately 15 in. long, and roughly or loosely bonded by circumferential twisting, oil application, or other means. There are extremely wide variations possible in denier, as governed by the speed of the take-up. Sliver fiber products enjoy wide usage in many fields and will be more elaborately discussed in Chapter 5.

PRODUCTION OF FIBER OPTIC ELEMENTS

Reverting almost to the original methods for producing glass mono-filaments from heated rods, this process, with substantial improvements, has become important again for producing fibers for visual fiber-optic and fiber electro-optic applications. The main differences are that glass compositions of extremely high purity are necessary for successful use of the ultimately produced filament. The compositions are established, the material is fused into a cylindrical "buhl" or preform and a filament or group of filaments drawn down under very carefully controlled conditions. Sophisticated methods are employed for producing jacketed or clad glass preforms (two or more glass compositions), and higher temperature compositions such as silica-quartz are drawn using a laser process.[47] See Fig. 4-9.

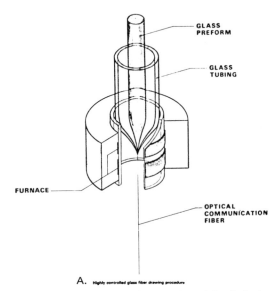

A. Highly controlled glass fiber drawing procedure

(*A*) *Single fiber.* To create the condition in which a "cladding" glass of lower index of refraction surrounds the "core," or light-carrying glass, glass tubing and an internally centered glass rod are indexed into a suitable furnace, and the composite fiber drawn away at high speed as the correct softening temperature is reached. A vacuum may be applied to aid in eliminating interface inconsistencies and defects. This is the most conventional method for producing fiber optics, since it can be precisely controlled. Obviously however, it is not the most economical.

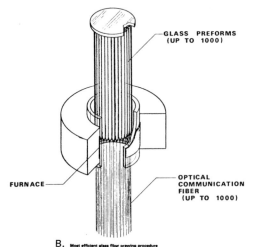

B. Most efficient glass fiber drawing procedure

(*B*) *Multiple fibers.* As many as 1000 fibers can be drawn simultaneously from previously prepared preforms (two or more glasses as core and cladding). This is one of the most economical methods of producing fiber optics, but has certain quality limitations. Some deterioration results from the second heating, and it is difficult to control the uniformity of furnace heat across the entire face of 1000 preforms. The most serious finished fiber defects are out-of-roundness and nonuniformity of filament diameter dimensions.

Fig. 4-9.   Six processes for drawing optical fibers. (*Diagrams A to E inclusive: Manufacturing Process for Galite® Optical Communication Fiber and Cable at Galileo Electro-Optics Corporation, Sturbridge, Mass., Re diagram E, see also German Patent No. 24 34 717. Diagram F, German Patent No. 24 15 052.*)

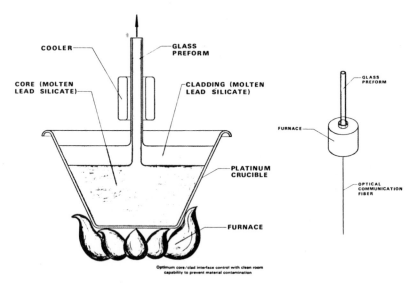

(C) The stratified melt process. Almost complete elimination of dirt, contamination, and voids or improper fusion at the interface (major causes of light loss) are possible, and are practically eliminated by floating the cladding glass on top of the core glass and up-drawing at a controlled rate. A preform for subsequent redrawing is formed in this case. However, in another adaptation, core and cladding are melted in a dual-compartment furnace, and a finished, clad fiber is drawn down vertically at high speed directly from the double melt.

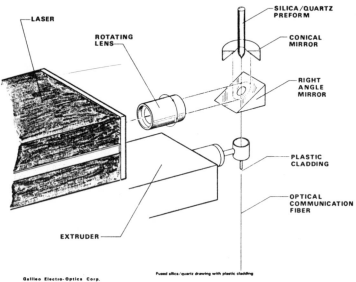

(D) Laser-drawing process. This method permits attainment of temperatures in excess of 3300° F to be concentrated at the locus of fiber attenuation from the preform. Although an expensive method, the drawing of pure silica/quartz optical fibers makes possible extended fiber-length transmission, and also transmission in ultraviolet wavelengths (see also Fig. 5-60).

Fig. 4-9. *(Continued)*

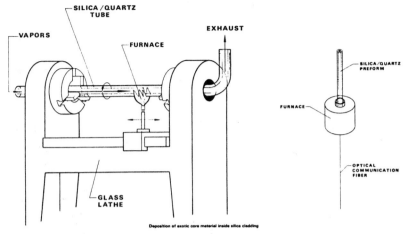

Deposition of exotic core material inside silica cladding

(*E*) *Vapor oxidation process.* In this method, chemical gases which produce pure SiO_2 and associated forms are passed through the heated silica/quartz tube and become deposited on the inside surface. Following deposition, the temperature is increased and the tube is necked down to produce a finite-size preform (no center void). This preform is subsequently reheated in another furnace and attenuated into an optical fiber. This is the most expensive optical fiber-drawing process, but costs are justified by the extreme purity of core-material gases, and control of contaminants in the closed system down to one part per billion (see also Fig. 5-61). In an important variation of this process, the composition of the gases may be sequentially varied to produce a fiber possessing a graduated index of refraction in the cladding material. The importance of this will be elucidated in the discussion of Fiber Optics in Communications.

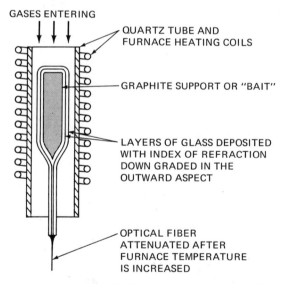

(*F*) *Use of a bait to condense glasses.* A cylindrical graphite support with a conical lower end is placed coaxially within a quartz tube. Layers of quartz glass are deposited on the graphite bait by passing mixture of SiH_2 and O_2 gases through the furnace. The gases are doped with various ingredients so that indices of refraction of the layers of glass sequentially deposited are graded lower toward the outside. Following deposition, the furnace heat is doubled and the graded-index optical fiber is drawn away vertically using a high-speed winding mechanism. In this method, no internal stresses, striations, or lines of demarcation are formed between glass compositions, and a fiber with perfectly graded index of refraction is obtained.

Fig. 4-9. (*Continued*)

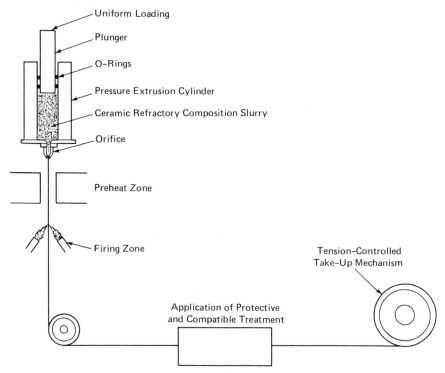

Fig. 4-10. Schematic diagram showing method for production of glassy mono-filament from a refractory fiber composition. (see ref. 48.) (*Courtesy SPI RP/C Institute*.)

EXTRUSION–FUSION METHOD

A method for production of a glassy monofilament starting with a refractory ceramic composition was developed by IIT researchers.[48] The mechanical components are illustrated schematically in Fig. 4-10. Obviously, the reduction of a ceramic body and subsequent fusion into the glassy state would necessarily have to proceed at a pace slow enough to preclude this from ever being a viable mass production technique.

Methods of depositing either single-crystal whiskers or vapor-phase material onto a metal fiber substrate to form a continuous filament are also described in the same paper. These fibrous products are usually crystalline and not glassy, however.

Chapter 5 | *Applications of Continuous- Filament Fiber Glass Products*

INTRODUCTION: CHARACTERISTICS AND TESTING

In Fig. 5-1 is illustrated an SEM photomicrograph of a bundle of continuous E-glass filaments which were drawn from a direct-melt furnace using a standard large-sized textile fiber glass bushing. In contrast with the disuniformity of the blown insulation-type glass fibers (Figs. 3-1, 2, and 3), the regularity of diameter and uniformity of the individual continuous filaments is quite obvious. The glob-shaped buildups (added-on portions) are the cured organic binder or sizing material applied for (1) lubrication, (2) adhering filaments into a strand, and (3) compatibility with the end-use medium, since almost all continuous-filament type products are considered as intermediates. Very recently, sizings have been developed which provide more complete and uniform coverage of the filaments.

Glass Compositions

Five predominant glass compositions are used in production of continuous-filament fiber glass products. These are presented in Table 5-1, and may be briefly described as follows: (1) The A type, a soda-lime glass, was the first used and still is retained in a few minor and noncritical applications; (2) the E- or electrical type, a borosilicate "cousin" of the early Pyrex® compositions, was developed to provide better resistance than A-glass to attack by water and mild chemical concentrations because of the increased surface area of glass in the fibrous state; (3) C- or chemical glass possesses considerably improved durability on exposure to acids and alkalis than does E-glass,[49] and is satisfactory for use in lead-acid batteries and also

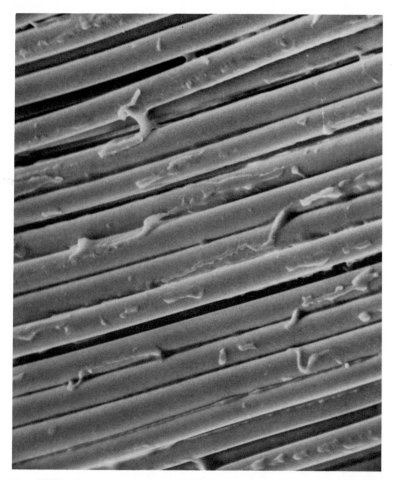

Fig. 5-1. SEM photomicrograph, 650X, of a continuous-filament fiber glass strand drawn from a large-size direct-melt bushing. Note the uniformity in filament sizes and parallelism. The extra deposited material is the adhesive-type organic binder applied for bonding, lubrication, and compatibility. (*Courtesy Johns-Manville Sales Corp.*)

laminated for the contact or interface exposure layer in RP/C corrosion structures;[50] (4) AR- or alkali-resistant glass is a comparatively new entity, and is employed in fibrous reinforcement of cement and concrete, which are both fairly new applications; (5) S-glass for high-performance applications mostly involving aircraft and aerospace possesses both tensile strength and tensile elastic modulus greater

TABLE 5-1. Glass Compositions Used in Major Applications of Continuous-Filament Fiber Glass.

Designation	A	E	C	AR	S
Characteristic	Common soda lime type	Electrical fiber comp. (used in most general-purpose RP/C)	Chemical glass (used in surfacing mats, etc., for corrosion resistance)	Alkali resistant (for reinforcement of concrete)	High-strength High-modulus (for high-performance structures)
SiO_2	72.0	54.3	64.6	60.9	65.0
$Al_2O_3 + Fe_2O_3$	0.6	15.2	4.1	0.27	25.0
CaO	10.0	17.3	13.4	4.8	—
MgO	—	4.7	3.3	0.1	10.0
Na_2O	14.0	$\{$ 0.6	7.9	14.3	—
K_2O	—		1.7	2.7	—
B_2O_3	—	8.0	4.7	—	—
BaO	—	—	0.9	—	—
TiO_2	—	—	—	6.5	—
ZrO_2	—	—	—	10.2	—
SO_3	0.7	—	—	0.2	—
As_2O_5	tr.	—	—	—	—
F_2	—	0.1	tr.	—	—

than E-glass by 33 and 20%, respectively, providing higher laminate strength-to-weight ratios, high strength retention at elevated temperatures, and a high fatigue limit.

Other glass compositions developed for fiberizing into continuous-filament products include several additional high-tensile and high-modulus glasses, for stronger and stiffer RP/C laminates; a high-lead glass for absorption of X-rays and other radiation; and a fiberizable glass with higher dielectric strength for improved properties in electronic applications.

Chemical and Physical Testing

Chemical and physical testing for the control of raw materials and finished glass composition include the following: (1) quantitative wet gravimetric chemical analysis (or reliable equivalent) of all incoming raw materials and of the melted glass; (2) control of all furnace operating parameters which bear on success of the melt; (3) petrographic microscopic examination and identification of abnormalities and anomalous deviations from the proper glassy phase (usually crystalline minerals) which sometimes occur in furnace and bushings and which generally interfere with fiber production; (4) continuous ongoing testing of physical behavior of the glass by a series of tests which include reboil (presence of dissolved gases in the glass), softening, annealing and strain points (ASTM methods C-338-57 and C-336-69), glass density (Preston Laboratories sink-float method), and flow rate, or amount of fiber accumulated on a winding drum through a single-tip test bushing at several temperatures near the production fiberization point (a measure of melt viscosity vs composition).

The most important parameter of continuous-filament fiber glass, akin to any other textile product, is its length-weight relationship. Systems most widely used in defining glass fiber "textiles" are yards per pound and its inverse, the Tex method. As stated, arbitrary groupings of practical filament diameters were set shortly following development of the 100 and 200 (204) hole bushings. This classified arrangement has persisted, and all interrelated functions are listed in Table 5-2. Definitions and relationships are also included.

Fiber strand length-to-weight relationship (yards per pound or Tex) is determined by use of a mechanical wrap reel, which draws a given length out of a textile package (6 to 120 yd for coarse to fine yarns or strands, respectively), after which the skein gathered is weighed. Figure 5-2 shows test in progress for determining (A) the length-weight value for a single glass strand from a forming package and (B) the same for a heavy staple fiber or sliver strand.

In addition to the scanning electron and petrographic (polarizing) microscopes previously referred to, the projection and comparator microscopes are useful for studying and controlling fiber glass textiles. The projection microscope (Fig. 5-3, Part A) permits enlargement and accurate measurement of filaments to 1000 diameters, and hence rapid determination of fiber class or "cut" and coordination with data on length-weight relationship. In Part B is shown a comparator microscope, capable of magnification of up to 50 diameters, and with a much larger field. It is useful in statistical determination of cut or broken fiber lengths, resulting in certain types of molding, and associated studies. Another useful measurement is the number of filaments per strand, and this may be carried out (however tediously) using a bacteria counter or similar laboratory equipment.

TABLE 5-2. Strand-Letter Designations with Corresponding Filament Diameters and Length-Weight Relationships for Multifilament Strands.

Continuous Fil. Ltr. Desig.	Filament Diameter (μ) Min	Max	Filament Diameter (in.) Min	Max	Percent Variation	200 Filament Strand Yd/Lb Max	Min	Tex Min	Max	2000 Filament Strand Yd/Lb Max	Min	Tex Min	Max
B	2.51	3.81	.00010	.00015	±20.0	88900	83850	2.6	5.8	18890	8385	26.2	59.0
C	3.81	5.08	.00015	.00020	±14.3	83850	47200	5.8	10.5	8385	4717	59.0	104.8
D	5.08	6.35	.00020	.00025	±11.1	47200	30200	10.5	16.4	4717	3019	104.80	163.75
E	6.35	7.62	.00025	.00030	±9.1	30200	20980	16.4	23.6	3019	2096	163.75	235.81
F	7.62	8.89	.00030	.00035	±7.7	20980	15420	23.6	32.1	2096	1540	235.81	320.96
G	8.89	10.12	.00035	.00040	±6.7	15420	11810	32.1	41.9	1540	1179	320.96	419.22
H	10.12	11.43	.00040	.00045	±5.9	11810	9700	41.9	51.0	1179	931	419.22	530.57
J	11.43	12.70	.00045	.00050	±5.3	9700	7500	51.0	65.5	970	754	511	655.03
K	12.70	13.97	.00050	.00055	±4.8	7500	6240	65.5	79.2	754	623	655.03	792.59
L	13.97	15.24	.00055	.00060	±4.3	6240	5240	79.2	94.4	623	524	792.59	943.25
M	15.24	16.51	.00060	.00065	±4.0	5240	4470	94.4	110.6	524	446	943.25	1107.01
N	16.51	17.78	.00065	.00070	±3.7	4470	3855	110.6	128.2	446	385	1107.01	1283.87
P	17.78	19.05	.00070	.00075	±3.4	3855	3360	128.2	147.2	385	335	1283.87	1473.83
Q	19.05	20.32	.00075	.00080	±3.2	3360	2950	147.2	167.6	335	294	1473.83	1676.89
R	20.32	21.59	.00080	.00085	±3.0	2950	2617	167.6	188.8	294	261	1676.89	1893.05
S	21.59	22.86	.00085	.00090	±2.9	2617	2333	188.8	212.8	261	232	1893.05	2122.32
T	22.86	24.13	.00090	.00095	±2.7	2333	2091	212.8	236.4	232	209	2122.32	2364.68
U	24.13	25.40	.00095	.00100	±2.6	2091	1889	236.4	261.6	209	188	2364.68	2620.14

Definitions and Conversion Factors:

1. Tex = Weight in grams per 1000 meters
2. Denier = Weight in grams per 9000 meters
3. Strand Count = Yards per pound/100
   (Example: K-fiber maximum yardage for 200 filament strand = 7500; strand count = 75.0)
4. Conversion yards per pound to Tex:
   Tex = 494420/Yd/Lb
5. Conversion denier to yards per pound:
   Yd Per Lb = 4,464,500/denier
6. Determination of yards per pound from filament diameter:
   $18.89 \times 10^{-4}/(\text{Filament diam in inches})^2$ = Yards per pound

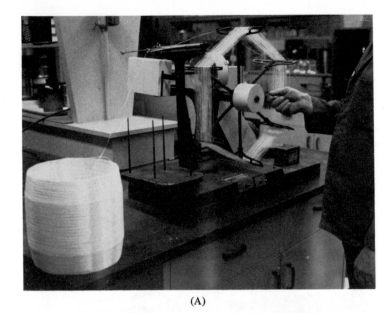

(A)

(B)

Fig. 5-2. Operation of a standard textile wrap-reel to determine length to strand weight relationship (yards per pound or Tex) of (Part A) a continuous-filament fiber glass strand, and (Part B) a heavier, bulky staple fiber or sliver strand. (*Courtesy Johns-Manville Sales Corp.*)

(A)

(B)

Fig. 5-3. Part A. Projection-type microscope for enlargement and measurement of fiber glass filament diameters up to 1000X. Part B. A comparator microscope useful for statistical determination of chopped or cut fiber lengths, aspect ratio, etc. (magnification up to 50X). (*Courtesy Johns-Manville Sales Corp.*)

Additional basic factors to be dealt with in controlling and evaluating fiber glass textiles are water content (both the capillary nature of the fine filaments, and the organic binder applied cause hygroscopicity) and ignition loss, or amount of binder or sizing solids applied (see Fig. 3-6).

Major application areas of continuous-filament (and related) fiber glass products to be described include the following: (1) reinforced plastics (RP/C), involving examples of all the various markets supplied; (2) fabrics, including all types, variations, and adaptations; (3) cordage, cabling, and sewing thread; (4) paper reinforcement; (5) wax and polymeric bonded strands; (6) ground cover; (7) abrasives; (8) reinforcement of cement, plasters, and gypsum; (9) staple fibers; (10) metal-coated fibers; (11) reinforcement of rubber and other elastomeric materials; and (12) fiber optics.

REINFORCED PLASTICS/COMPOSITES (RP/C)

Introduction

This portion of the fiber glass spectrum of products is also referred to as FRP (fiber glass reinforced plastics), RP, and Advanced Composites, the latter being also concerned with some other reinforcements additional to fiber glass. The RP/C designation was set forth by a specific activity group or "institute" within the Society of the Plastics Industry (SPI) to include the entire field. Personal preference or custom should dictate which terminology is used. The most salient criterion is that complete and universal understanding should be achieved.

In RP/C, fiber glass is mixed, impregnated, or saturated with a liquid, polymerizable resinous product which cures to a rigid solid by action of catalytic chemicals either at room temperature or under heat and pressure. Types of resins include thermosetting polyesters, epoxies, phenolics, vinyl-esters, furans, and all categories of thermoplastic resins. Many filler types are incorporated for various improvements.

Molding methods are almost legion, with new and vital improvements in methods and processing appearing annually. At present count, more than 22 general molding methods in six basic processing categories are widely used.[50]   These include hand-and-spray saturating and mixing (hand lay-up and spray-up), continuous impregnation and cure of flat or corrugated panels and of solid rod or hollow bar stock, compression, transfer, and injection molding (including presaturating BMC and SMC* techniques), filament wrapping or winding, centrifugal and static casting, rotational molding, cold forming, and many others which are indigenous to RP/C and/or combinations of those above.

The wide acceptance and rapid growth of RP/C as a material is the fortuitous result of the basic high strength of the fiber glass reinforcement and the excellent bonding power, processability, and chemical durability of the resins. Tensile filament strengths of E-glass approach 500,000 psi. The chemical composition of all types of resins used in RP/C may be varied to provide molding stability, high hot-strength, resistance to specific corrodants, fire retardance, and many other desirable properties.

RP/C is making inroads in many markets due to general properties of excellent molded surface finish, almost unlimited size, lightness of weight, insulation against heat transfer and electricity, and many other welcomed attributes. These will become evident in the ensuing discussion, built around the nine major markets for RP/C which include the following: (1) aircraft and aerospace; (2) appliances and equipment; (3) construction; (4) consumer goods; (5) corrosion-resistant products; (6) electrical rods, tubes, and components; (7) marine and marine accessories; (8) land transportation; and (9) miscellaneous, including equipment for materials handling, protective gear, farming, and industrial tooling.

It should be pointed out that, although much emphasis is given to processing methods in RP/C, an article or product from any molding method can be fabricated which will fill requirements or particular product specifications for any of the nine market categories.

In Fig. 5-4 are shown basic fiber glass intermediates used in RP/C and also some related types of reinforcement. Products used as reinforcement for plastics include roving, woven roving, chopped strand mat, chopped strands, and milled fibers. Special products combining chopped strand mats and/or woven roving with each other and with straight unidirectional roving are also widely used.[51]

Aircraft and Aerospace Market

Introductory to aircraft and aerospace is the story of a young teen-age hopeful who became interested in rocketry and space vehicles. With the help of some illustrated "science wonders" magazines and

---

*Bulk Molding Compound and Sheet Molding Compound for RP/C.

Fig. 5-4. Fiber glass products used in plastics reinforcement include (2) roving, (3) woven roving, (6) chopped strand mat, (7) chopped strands, (8) milled fibers, (9) fiber glass twisted yarn. Other products shown are: (1) basic forming package, (4) serving package for paper reinforcement or electrical applications, (5) coil of phenolic-impregnated roving for use in flexible fiber glass insulating duct, (10) wax-bonded cord for electrical applications, (11) phenolic bonded industrial mat. (*Courtesy Johns-Manville Sales Corp.*)

his high school chemistry instructor, he learned to mix solid fuels which when ignited, would send small model rockets zooming skyward with a distinctive P-f-f-f-f-f-f-f-t.

Over the course of several months' experimentation, all fuel mixes and rockets did not perform in the same manner. Some were high flyers, while others took off with only partial success and a sort of half-hearted Pfft.

The boy's father watched these proceedings with great interest, and finally, seeing an opportunity for moral preachifying, said "Son—this should be a lesson to you; always put your best pf-f-f-f-f-tt forward."

Many RP/C components have been designed and built for aircraft application. Because of the rigidity required, reinforcements other

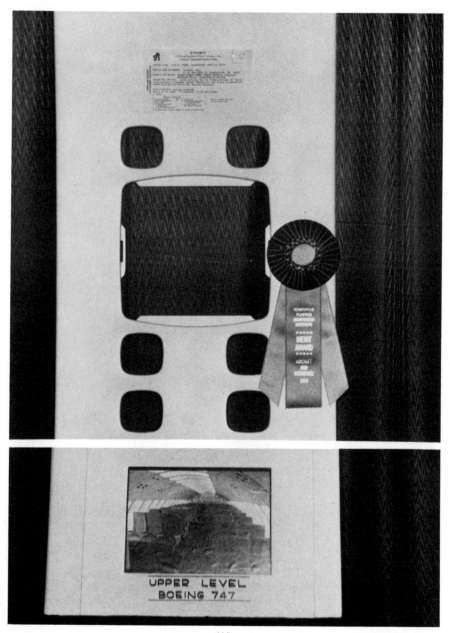

(A)

Fig. 5-5 Part A. Sectional overhead RP/C panels molded for use in upper-level passenger compartment of a Boeing 747. This part won an SPI RP/C Institute Exhibit Design Award at one of the recent Annual SPI Technical and Management Conferences. (*Courtesy The Boeing Co. and SPI RP/C Institute.*)

(B)

Fig. 5-5. (*continued*)

Part B illustrates 100% use of fiber glass RP/C in the exterior of a flight vehicle. This Fairchild "Goose" missile, described as a "long-range, air-breathing missile weapon system," was designed for support of Strategic Air Command missions. It was superseded by the "Whale" guided missile, which was also wasted on launch, but could be carried under the wing of a B-52 or other bomber plane, and fired only at the discretion of the gunner. (*Courtesy Fairchild Corp., Aircraft Division.*)

than glass are usually more frequently used for external parts of the aircraft structure. However, entire external fuselages of smaller planes have been constructed using glass fibers as the reinforcement. An excellent example of use of fiber glass RP/C in aircraft is in the interior of a commercial plane, the Boeing 747, shown in Fig. 5-5, Part A. Part B illustrates a guided missile, showing use of fiber glass RP/C for the entire exterior of a flight vehicle.

RP/C in Appliances and Equipment

Whereas fiber glass insulation is used to control heat loss and excessive noise in appliances and equipment, molded RP/C parts are used for housings and bases. Frequently, fiber glass or other insulation is encapsulated between RP/C skins. The unusual RP/C properties of high strength-weight ratio, excellent molded surfaces, dimensional stability, high dielectric strength, and parts consolidation have been responsible. Use of RP/C for frames, bases and housings for computers, time-share terminal units, calculators, typewriters, and television sets, plus air conditioning units and similar appliances is well established. One interesting application, illustrated

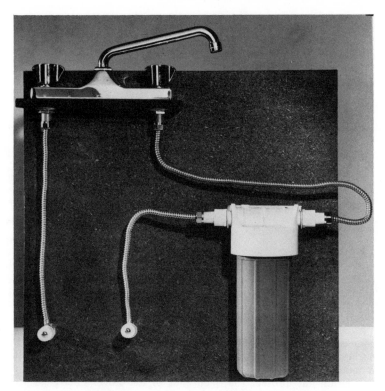

Fig. 5-6. Display model of an RP/C filter housing for use in improvement of domestic water supply. Both cap and housing are injection-molded using fiber glass reinforced thermoplastics. (*Courtesy Johns-Manville Sales Corp.*)

in Fig. 5-6, comprises a molded reinforced thermoplastic filter housing for improvement of domestic water supply.

Large equipment applications are exemplified by use of fiber glass RP/C in construction of the Alaska pipeline (see Fig. 5-7).

Construction

Numerous adaptations of fiber glass reinforced plastics in all types of construction have taken place. More and more activity has been generated in recent years based upon unique suitability of RP/C for building products. The list includes both interior and exterior building components for residential, commercial, industrial, and farm construction. Also important are the well-accepted tub and shower units, and other ancillary equipment such as patio covers and garage

(A)

Fig. 5-7. Use of molded fiber glass RP/C housings to cover structural support stations in construction of the Alaska pipeline. The fiber glass RP/C elliptical-shaped modules (Part B) (fabricated in two halves) encapsulate polyurethane foam insulation and protect the open ends of the blown fiber glass insulation which is adhered to the metal sheets, kerfed, and wrapped around the linear extensions of the pipeline using the mechanical enfolder (Part A). The "saddle" upon which the pipeline rests at each support station (directly beneath the modules) is also of RP/C (Part C). The part is non-load bearing and also houses foam insulation. A teflon coating is applied to the saddle to permit a "sliding" action of the pipeline, thereby accommodating shifting due to expansion and contraction.   "Bumper" modules which have flat sides, to contact and slide against the vertical support posts, are designed and installed at critical points such as bends, thus preserving the intended alignment of the pipeline when it undergoes linear expansion and contraction. The fiber glass insulation adhered to the sheet metal and enfolded around the pipeline sections is of higher density at the top so that when properly positioned, a maintenance man or inspector is able to walk on the top of the line without damaging the insulation. Oil enters the line at a temperature between 120 and 140°F, flows at the rate of 7mph, thereby requiring 4½ days to span the 800 miles from Prudhoe Bay to Valdez Terminal.   There will be 9.04 million barrels of oil in the line at full capacity.   The project cost approximately $7 billion.   Owens-Corning Fiberglas Corp. made an independent arctic fiber glass insulation and construction testing and development study which cost over $2 million. The information gained from this program enabled them to secure the insulation contract from ALYESKA. (*Courtesy Owens-Corning Fiberglas Corp.*)

(B)

(C)

Fig. 5-7. (*continued*)

doors. Reusable utility segments such as RP/C forms for casting or
pouring concrete enjoy wide acceptance.

Examples shown herein of applications of fiber glass RP/C in the
construction field include the following.   Figure 5-8 illustrates

(A)

Fig. 5-8. Part A. Exterior RP/C building panels provide the appearance and
servicability of concrete yet weigh only one-fourth as much.   Panels up to
8 × 24 ft are fabricated, and 200,000 sq ft of exterior wall surface area were
covered in this particular building construction. (*Courtesy Uni-Systems Corp.
and PPG Industries for the Detroit, Michigan Renaissance Center.*)   Part
B. Building fascia modular panels 3.42 m high × 2.75 m wide for clustering on
external structural surface. The window projection of 0.7 m is to help guard the
inside from the incidence of sunlight and glare. (*Courtesy City Government,
Papua, New Guinea, and Transfield Pty. Ltd, Sydney, Australia.*)

(B)

Fig. 5-8. (*continued*)

modular RP/C building units.   Part A shows erecting into position
exterior RP/C building panels which weigh only one-fourth as much
as concrete by virtue of adhering them to a structural steel frame
during fabrication. Panels up to 8 X 24 ft may be fabricated. Part
B shows building fascia panels with window emplacement hole and
provisions for mounting.   Figure 5-9 Part A shows exceptionally
attractive and practical residences in which the RP/C is formed into
a bent, modular unit permitting many combinations for interesting
home construction.   Part B shows corrugated translucent architectural paneling being installed as a patio cover. Additional uses for
this material include solar heating elements, garage doors, industrial
building skylights and sidelights, yard and swimming pool fences and

(A)

(B)

Fig. 5-9. Part A. Attractive residential homes constructed using prefabricated modular RP/C units.[52]   (*Courtesy Polyarch*™ *Homes, Rudkin-Wiley Corp.*) Part B. Installation of corrugated RP/C translucent architectural paneling as a patio roof. Other applications include solar heating panels, garage doors, fences, and awnings. (*Courtesy Owens-Corning Fiberglas Corp.*)

(A)

(B)

Fig. 5-10. Part A.   An RP/C bathtub fabricated with an imbedded or laminated marble design originally printed on glass cloth.   The molded finish possesses great depth and luster, and the part is extremely servicable, withstanding 500 hr exposure by repeated dipping into boiling water, and resisting a 50 ft lb impact. (*Courtesy Ina Seito Co. and SPI RP/C Institute.*)   Part B.   A four-piece high-speed compression molded RP/C residential or commercial tub and shower unit (*Courtesy Owens-Corning Fiberglas Corp.*)

awnings. Use of RP/C in bathroom fixtures is shown in Fig. 5-10. In Part A is presented a bathtub with an exceptional simulated, laminated-in onyx surface finish which possesses excellent resistance to mechanical impact and boiling water. In Part B is shown a four-piece compression-molded tub and shower unit with excellent surface finish and properties, plus the advantage of high production molding capacity.

Consumer Goods

Consumer products fabricated in RP/C are generally related to at-home leisure or recreational activities. Home products include basic frames for furniture as well as the finished items, both traditional

Fig. 5-11. Fiber glass skis being given a real opportunity to perform on a fast downhill run. High RP/C molded strength plus resiliency reduce ski chatter and improve control and manipulability, permitting the skier to increase his speed at least 15 mph to approximately 70 mph. (*Courtesy Owens-Corning Fiberglas Corp.*)

and modern. Also included are divider screens, decorative and utility trays, wall plaques, luggage, and other valuable household gadgetry.

Design and sales of recreational and sports equipment have been spurred by manufacturers' capability of supplying them in RP/C. The list of products includes outdoor and patio furniture, swimming pools, playground equipment, skis, golf clubs, portable tennis courts, tennis racquets, fishing rods, vaulting poles, snowmobiles, garden tractors, surfboards, and skateboards.

The controllable flexibility of fiber glass RP/C to produce varying stiffnesses required in vaulting poles, golf club shafts, etc., plus the combination of high mechanical strengths, lightness of weight, easy formability, durability, molded-in color and excellent surfaces, and resistance to corrosion and wear are all viable properties which favor use of RP/C material for consumer products.

Prime examples are shown in Figs. 5-10, 5-11, 5-12, and 5-13.

Fig. 5-12. Transported to the installation site via helicopter, this mountain climbers' retreat or "bivouac" is successfully resistant to stresses induced by winds up to 100 mph, snow, and rapid temperature changes. Foam-core insulation with inner and outer RP/C skins makes it a comfortable and convenient shelter for 12 to 14 persons at one time. (*Courtesy Vetrotex International, Division of Saint-Gobain Industries, Jawoplast-Karnten in St. Andrae im Laval (Austria), and Architekt Ohnmacht.*)

Fig. 5-13. Ready moldability to a designer's extreme aesthetic demands plus excellent surface finish make RP/C a most suitable and adaptable medium for these modern furniture pieces. (*Courtesy SAVID, S.P.A. Como-Tavernola, Italy.*)

Fig. 5-13. (*continued*)

Corrosion-Resistant RP/C Products and Equipment

Exhaustive analyses of the mechanisms of corrosion in fiber glass RP/C structures are available in the literature, together with recommended construction practices for optimum field and exposure performance.[50,53,54] Outstanding research, development, testing, and technical service have been carried out by the chemical companies which manufacture polyester and associated thermoset and thermoplastic laminating resins. Also, as previously indicated, considerable emphasis by fiber glass manufacturers has been placed on suitable fibrous glass compositions for specific durabilities, and also development of the most functional organic binders and sizings to promote best adhesion and resistance on exposure to a wide range of chemicals and concentrations. The SPI RP/C Institute Annual Proceedings volumes are replete with research papers correlating resin types with chemical durability data, together with case histories and performance studies of fabricated RP/C end products.

Resin types involved include general-purpose polyesters and epoxies, isophthalic, bisphenol A, and vinyl-ester polyester types, chlorendic anhydride, brominated and filled types which are fire retardant as well as chemically resistant. Also included are the furfuryl alcohol (furan) resins, which possess excellent chemical durability, and have been made much more workable in the light of recent improvements in handling and catalysis.

Information and data regarding resistance of RP/C structures to any specific chemical in any concentration may be readily elicited from any of the several large resin producers possessing reputable expertise in the field of chemical resistant RP/C.

A word of caution is apropos: any contemplated RP/C system or design should be thoroughly checked out prior to installation for both mechanical and chemical stability, and structural competency. There is no excuse for material misuse from which unfortunate personal injury or extensive property damage might result. Data on stress-corrosion testing is an important factor, and should be sought out rather than accepting separately derived mechanical and chemical test information.

Engineers and managers have realized that superior corrosion-resistant properties exist in RP/C as compared to traditional materials. Installations of equipment providing high-quality performance include the following: ducts, fume hoods, processing and storage tanks, piping up to 12 ft diameter produced on a continuous machine, floor grating, ladders and railings, cable trays, small and large cooling towers, components for water and sewage treatment facilities, sewage flow meters, and many other applications.

Applications exemplifying the eminent success of RP/C in the corrosion-resistant market are the manhole housing and sewage piping sections (Fig. 5-14) and the huge outdoor tanks for storage of wine (Fig. 5-15).

(A)

Fig. 5-14. Part A. Installation of a manhole housing to provide access to a sewer line. The RP/C part is 4 ft in diameter and may be made in lengths up to 35 ft. These components are highly resistant to sewage liquors, sewer gases, and associated chemicals including salt or salt water, corrosive ground or soil conditions, electrolysis and/or stray subsurface electrical currents. (*Courtesy Owens-Corning Fiberglas Corp.*) Part B. RP/C corrosion-resistant pipe with flanged fittings which is being used to supersede cast iron pipe and fittings in a sewage treatment plant in Launceton, Tasmania. (*Courtesy Transfield Pty, Ltd., FRP Division, Sydney, Australia.*)

(B)

Fig. 5-14. (*continued*)

Fig. 5-15. A tank farm of exceptionally large corrosion-resistant RP/C vessels constructed for the storage of wine. (*Courtesy of VITROFIL S.p.A., an affiliate Company of MONTEDISON Group, Milano, Italy.*)

Electrical and Electronics Market

Substantial incursion of RP/C into fields requiring desirable properties for electrical and electronic applications such as weather stability, high dielectric strength, high arc resistance, and good mechanical toughness was rapid once these materials were proven in a few critical areas. Such uses have resulted as distribution-pole hardware, switchgear, transformers, telephone equipment, printed circuit boards and parts (see Fabrics), computer parts and housings, and many other components. Size extremes of useful electrical parts range from coin size to 55 ft diameter radomes (see also RP/C Appliances and Equipment).

A fortuitous circumstance also exists for use of RP/C in electrical and electronic applications: fabrication by almost every known RP/C processing method provides parts that satisfy the many varied requirements of the entire electrical field.

Figure 5-16 Part A shows an assortment of molded elements, housings, and insulators all relegated to the electrical and electronics industries. Part B shows a well-designed and highly functional electric circuit breaker. Part C illustrates use of fiber glass RP/C in equipment requiring high dielectric strength and excellent arc resistance. These are line work booms and man-carrying buckets for bare-hand work on high voltage, telephone, television, and other lines and cabling installations.

Marine Products and Marine Accessories

Properties of "fiber glass" (the boat manufacturers', sellers', and users' synonym for reinforced plastics of marine craft and vessels of all descriptions) are so favorable that approximately 70% of all outboard pleasure boats 15 ft and longer are now constructed using the material.

The major benefits of using RP/C in any boat construction are moldability to almost any boat design or size, seamless construction, high strength and great durability, minimum maintenance, and freedom from corrosion, rust, dry rot, and water logging.

In addition to small motorcraft and a host of small recreational water-sports surf and ski boats, sailboats of all sizes and descriptions are being fabricated using RP/C. Also fabricated are commercial and military hulls, including the following: fishing boats, lighters (LASH) (saving 40 tons per unit compared to steel construction), submersibles, hovercraft for fast commuter service, plus both 120 ft American Board of Shipping research vessels and 153 ft British minehunters.

Also employing considerable annual tonnage of composite fiber glass, resin, and filler product are the diverse marine accessories, a partial list of which would include: sealed pontoons for floating docks and swim floats, outboard motor shrouds, submarine fair-

(A)

Fig. 5-16. Part A. An assemblage of various molded RP/C parts for use in the electrical and electronic industries. Please note that many have metal inserts molded in (most all are threaded) for mounting and/or bolting with other components for the final assembly. (*Courtesy Glastic Corp.*) Part B. A circuit breaker in which the entire housing and most of the interior functional parts are molded in high-dielectric, high arc-resistant fiber glass polyester which also possesses dimensional stability and weather resistance. (*Courtesy Westinghouse Corp. and SPI RP/C Institute.*) Part C. Man-carrying RP/C buckets and extensible booms for working on high-tension telephone and television cables and lines. The men are so well insulated electrically, that direct work on live high-tension wires (bare-handed method) is possible. Also, much time is saved in not having to climb poles, not hoisting gear with other crane equipment, etc. (*Courtesy Owens-Corning Fiberglas Corp.*)

(B)

(C)

Fig. 5-16. (*continued*)

(A)

(B)

Fig. 5-17. Part A. The large 153 ft all-fiber glass RP/C mine hunter, HMS "Wilton", built and commissioned into the British Royal Navy. This ship was the mainstay in a combined British, Egyptian, and American operation to clear the Suez Canal of World War II debris including crashed aircraft, rockets, unexploded bombs and mines, and sunken vessels containing ammunition, thus again normalizing commercial shipping for a large number of the world's trading nations. (*British official photograph, Courtesy Central Office of Information, London, England.*)   Part B. Photo shows a 75 ft fiber glass RP/C Japanese fishing boat being launched. This boat displaces 99 tons, and vessels of this type are used to a great degree in Japan. (*Courtesy Nippon Glass Fibers Co. Ltd., Tsu, Mie Pref., Japan.*)

waters, tanks of all descriptions (fuel and storage), ladders and rails, seating, paddles, spars, and many other items.

Classic examples of use of the best qualities of fiber glass RP/C in marine applications are the following: the above-mentioned 153 ft British Royal Navy's all-fiber glass RP/C mine hunter, HMS

(A)

(B)

Fig. 5-18. Part A. Fuller's rowing needles, a conceptual design by Bucky Fuller for a single-scull sports craft. The twin needles weigh only $11\frac{3}{4}$ lb each, making possible a low total craft weight of 42 lb. They are 21 ft long, 8 in. beam and $5\frac{1}{2}$ in. deep, providing such minimal surface contact with the water that a 225 lb man can propel the twin shells at speeds up to 15 mph. The needles are of twin-skin sandwich construction, promoting lightness of weight and rigidity. (*Courtesy W. E. Hunnicutt, Big Bend, Wis., and SPI RP/C Institute.*) Part B. Illustrates a one-man white-water sports canoe or kayak. The unusual properties of RP/C make this light, rigid, and rugged construction possible. (*Courtesy Hyperform, Hingham, Mass.*)

"Wilton", used to help clear the Suez Canal of unexploded bombs, mines, etc., remaining from World War II, and a 75–80 ft Japanese fishing boat (displacing 99 tons) (see Fig. 5-17, Parts A and B). Also shown are a rowing or sculling needle and a one-man "white water" canoe or kayak (see Fig. 5–18, Parts A and B).[55]

Land Transportation

The first real success-splash of fiber glass RP/C in the automotive or land transportation field was the 1953 Corvette which had a complete fiber glass body. While this vehicle is still in production and will always be made with a 100% fiber glass body, merchandising emphasis has shifted over the years to placement of many singular reinforced plastic components wherever they will satisfy the stringent automotive performance requirements.

At the most recent count, 154 different automotive applications in RP/C existed, not to mention truck, motor home, and trailer parts. Of these, 86 are functional and not strictly cosmetic (or used only for surfacing and appearance).

The prime contribution of RP/C is weight saving, and this factor alone will go a long way toward helping auto makers satisfy future Federal energy consumption requirements for cars of 20 miles/gal. by 1980 and 27.5 miles/gal. by 1985. Use of more and more RP/C components will permit cars to meet these requirements by contributing lightness and still maintain some semblance of roominess and comfort. Whereas 160 total pounds of plastics (3.7% of curb weight) were built into cars in the 1975–77 era, predictions are for 250 lb (7.1% CW) in 1980 and 350 lb (10% CW) in 1985. This is energy effectiveness at its best!

Other desirable properties contributed by fiber glass RP/C to automotive and land transportation vehicles and equipment are the following: (1) freedom from rust and corrosion and 100% elimination of other problems contributed by metals (the first original RP/C Corvette body off the molded production line from 1953 is still in excellent uncorroded shape and still mobile); (2) parts consolidation, or elimination of many separately formed and welded, bonded, or assembled components as is required for steel body construction (truck hoods and motor homes are two prime examples of successful parts consolidation using RP/C); (3) lower tooling costs and great facility of production moldability. (This is not a process-oriented literary effort, but the development and availability of new fiber glass sheet molding compound (SMC) and associated methods of molding have provided a real upward surge in the practicability of use of RP/C in automotive applications. One set of dies is required for producing an RP/C part, while metal parts production require a minimum of 12 massive successively staged stamping dies, hence considerably larger capital expenditure.)

Figure 5-19 illustrates an assemblage of a great multiplicity of parts for automotive uses. Easily recognizable are front ends, engine hood covers, fender extensions, hood scoops, lamp housings, air-cleaner housings, window inserts and frames, dash instrument panels and accessories, consoles, interior trim components, and others. Also manufactured but not shown are rear-end panels, fender liners, rear wagon doors, air conditioner housings, package trays, retainer panels, and wheel-opening covers.

The functional automotive parts fabricated using fiber glass RP/C

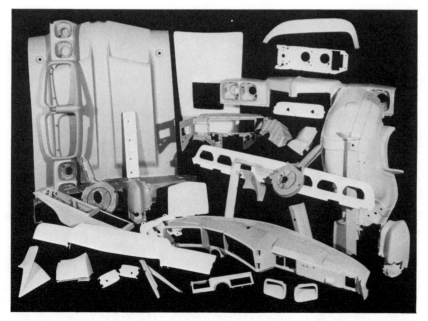

Fig. 5-19. A few of the more than 150 separate RP/C parts used in more than 37 different models of cars. The total number is growing rapidly. (*Courtesy Owens-Corning Fiberglas Corp.*)

include rods, valves, ducts, brackets, brake pistons, gear trains, distributor rotors, fuel pumps, gearshift housings, hydraulic pistons, gears, washers, valve plates, fans, latches, pulleys, oil seals, electrical junction boxes, and others. Structural members in RP/C such as transmission support bars, radiator support housings and front bumper back-up beams have been in production for several model years. An experimental Pontiac Phoenix car with all-RP/C wheels plus other functional and support components was shown at a recent SPI RP/C Institute Annual Conference and Exhibit.

An automotive door made using RP/C is under engineering study; it would weigh only 37 lb versus 66 lb for an equivalent steel door. Use of fiber glass reinforced elastomeric materials (tires and bumpers) is also in the ascendency (see Reinforced Rubber and Elastomers).

Additional examples of use of RP/C in automotive and related land transportation applications include the following: Figure 5-20 shows a Daimler-Benz test car effective because of its special rigid and

Fig. 5-20. Daimler-Benz test sports car recently exhibited in Dusseldorf, West Germany. Both the self-supporting chassis and fold-open doors are fabricated using a special type moldable RP/C sandwich construction to provide rigidity plus lightness of weight. (*Courtesy BAYER AG, Leverkusen, West Germany.*)

Fig. 5-21. A one-piece compression-molded, forward-tilting engine-hood and fender combination for a large truck cab. Parts consolidation and ruggedness permits savings in both labor and assembly time. Also the lightness of weight represents fuel energy savings and hence will permit hauling of heavier payloads. (*Reproduced by permission GMC Truck and Coach Division, General Motors Corp.*)

lightweight sandwich-type construction; Figure 5-21 shows a one-piece, forward-tilting engine and fender hood for a truck; Figure 5-22 Part A shows RP/C refrigerator-type railway cars in an assembly line, and Part B shows highway guard rails.

Miscellaneous Product Market (Specialty Products)

The tooling industry is the largest single facet making up this final marketing category for RP/C parts. A comprehensive list would include such highly servicable and behind-the-scenes components as temporary or short-run forming dies for plastics and metals, checking fixtures, hydroforming shapes, hammer forms, master models, stretch dies, mass-cast forms, foundry patterns, and many other creative and utilitarian elements.

Other examples are industrial equipment such as hard hats, welding masks and shields, shipping racks, utility trays and containers, and parts-handling systems. Hospital and store conveyor systems (electronically controlled) and farm equipment (tractors, housings, tanks,

(A)

(B)

Fig. 5-22. Part A.   Railway refrigerator cars in an assembly line. Construction is RP/C skins with polyurethane foam-core. (*Courtesy VITROFIL S.p.A, an affiliate company of MONTEDISON Group, Milano, Italy.*) Part B. Highway guard rail fabricated using resilient RP/C to absorb impacts and made slightly thicker to withstand stresses developed. (*Courtesy Nippon Glass Fiber Co., Ltd., Tsu, Mie Pref., Japan.*)

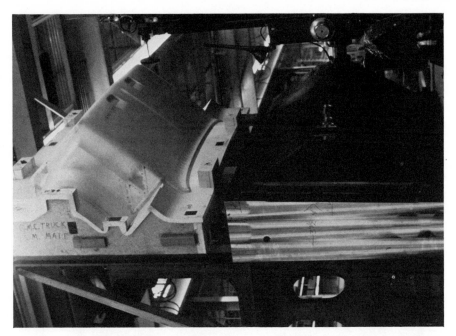

Fig. 5-23. Use of fiber glass RP/C (structure on left) as the model tool in an automatic duplicating machine. The "plug" fiber glass part is cast out of a "cavity" type mold (also fiber glass) previously fabricated from the original wood pattern. The rigid, unyielding, true-detail fiber glass surface provides an excellent base from which the duplicating head will trace and machine an identical contour into the cast tool-steel shape on the right. This will form the "plug" portion of a production mold for reinforced plastic SMC parts. The operation depicted here is machining the mold for the GMC truck engine cover shown in Fig. 5-21. (*Courtesy Modern Tools, Div. Libbey-Owens-Ford Co.*)

troughs, etc.) are also included. Figure 5-23 illustrates one of the more recent adaptations of RP/C in the tooling industry.

FIBER GLASS FABRICS

Introduction

This discussion of fabrics will comprise (1) all regular woven types including tapes; (2) braiding; (3) knitted fiber glass fabrics; (4) scrim (actually a nonwoven but clothlike material); and (5) fabrics produced from sliver.

The weaving operation is performed on many varying types of looms, and glass fabrics are adaptable to essentially each and every one. Weave types include the following: (1) plain, (2) basket, (3) twill, (4) satin, and (5) leno. Each has basic properties varying between (1) stability and mobility (or sleaziness) of weave; (2) flatness or firmness and drape or deep-draw; and (3) strength, which varies in weave direction and cross-weave direction. Other woven glass fabric variables are (1) thickness limit = 0.001 to 0.050 in.; (2) construction, or number of yarn ends per inch in the weaving direction (warp) and cross-weaving direction (fill) (may vary between over 100 for exceptionally fine-layered yarns to less than 3 for wide, coarse, or untwisted and unplied yarns such as woven roving); and (3) weight (1 oz/sq yd to over 16 oz/sq yd); (4) fabric strengths are an additional variable and range between 35 and 1400 lb/in. direct tensile pull (clamped). Figure 5-24 illustrates several types of fabrics and tapes, and yarn package types used in weaving. Heavier cordage types are also shown (see Yarns).

Tapes are usually prepared in a plain weave. Single-head or multiple-band looms are used in fabrication of tapes. Braiding is

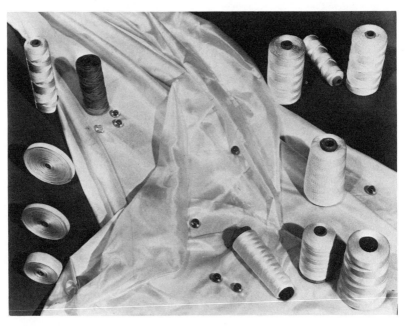

Fig. 5-24. Several types of fiber glass fabrics, tapes, and several of the types of yarns used in their preparation. Heavier cordage products are shown at the top of the photo. (*Courtesy Owens-Corning Fiberglas Corp.*)

Fig. 5-25.   Fiber glass yarn on suitable packages for delivery being braided into a tubular woven product, referred to as "sleeving." (*Courtesy Owens-Corning Fiberglas Corp.*)

performed on special carousel-type equipment (see Fig. 5-25) to produce tubular woven products.

Yarns of course constitute the basic elements of woven structures, and the possibilities for variations in the type and size are almost infinite. Filament diameter and number of filaments per strand may be varied, and also the number of twists per inch, usually varying between 0.5 and 10.0 (tpi). Direction of twist is designated as S or Z (right-or left-hand) and the ply direction is always the reverse or opposite of the twist direction, thereby providing yarn balance.

Addition of twist or twist-and-ply to fiber glass yarns is necessary to assist the yarn in withstanding the rigors of weaving, but basic sizings or binders, usually a starch type, must also be added to ease the yarn past mechanical contact points and prevent excess filamentation. Water sprays to create high humidity are also a great adjunct to loom processing of fiber glass.

Likewise, many different treatments to the finished fabric are available. First it is usually required to remove by heat-cleaning the original starch yarn sizing and subsequently apply a compatible finish using a spraying, dipping, slashing, and drying or other procedure. The finishes render the fabric compatible with epoxy, melamine, phenolic or polyester resins, or simply add abrasion resistance or other protection against mechanical degradation if the fabric is to be used unlaminated.

Scrim (nonwoven) fabrics are prepared by adhering yarn at crossover points to form plain or square patterns in which the distance between contact points varies from $\frac{1}{4}$ to 4 in. or more as specified. The adherents may be asphalt or like thermoplastic material with a low temperature melt index.

Fabrics made from sliver yarn are usually either plain weave or high-modulus weave (coarse yarn in warp, fine yarn in fill, or the opposite). The loose, noncontinuous nature of the sliver or staple fiber strand is distinctly evident in Fig. 5-26.

In addition to the tensile and other test parameters mentioned above, finished fabrics may be evaluated according to the following: (1) Mullen burst strength; (2) tear strength; (3) air permeability; (4) dielectric strength and constant, loss tangent and water absorption for efficacy in electrical applications; (5) endurance tests including MIT flex fold, Stoll flex, heat flex, and Taber abrasion.

Applicable military specifications for cloth and end-use performance factors are: HH-C-466B, MIL-Y-1140C, MIL-C-7514B, MIL-R-7575B, MIL-P-8013C, MIL-C-9084B, MIL-R-9299B, MIL-R-9300A, MIL-C-10797A, MIL-I-17205B, and MIL-C-20079D.

The various applications and end uses of fiber glass fabrics are described in the ensuing portion and include the following: (1) printed circuit boards and other reinforced plastics; (2) orthopedic casts; (3) paper reinforcement; (4) coarse industrial fabrics; (5) decorative fabrics, wall covering, and clothing; (6) screening; (7) filtration; (8) high temperature fabrics; and (9) coated fabrics.

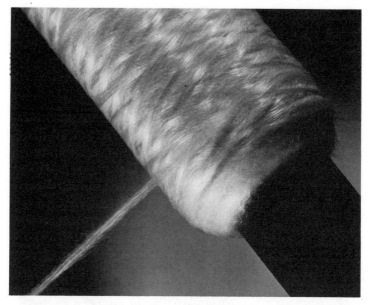

Fig. 5-26. A detailed view of the configuration of a fiber glass bulk fiber or sliver strand, and the type package upon which it is produced. (*Courtesy Owens-Corning Fiberglas Corp.*)

Printed Circuit Boards and Other Electrical and RP/C Applications

The largest single area of usage of fiber glass fabrics is in printed circuit boards (pcb's) laminated to desired thickness usually with epoxy resins, for electrical and electronic gear. Although many other materials, even other forms of fiber glass (mats, etc.) have been evaluated, the performance of fabric remains superior due to uniformity, ease of handleability in mass production continuous throughput laminating operations, punchability, and many other factors. Figure 5-27 shows a typical pcb printed with the metallic coating and etched to reveal by discovery the ultimate circuitry. Some of the larger electrical components eliminated by use of pcb's are pictured in the background.

In many other electrical applications, glass fabrics are used because of their excellent dimensional stability and superior electrical insulating properties. In Fig. 5-28 Part A is shown an operator winding fiber glass tape over heavy connecting leads for the magnets of a large electric motor or generator. These parts are varnished for final

Fig. 5-27. Shows the orderly array possible with a pcb made using glass fabric. Some of the older, larger-sized electrical elements eliminated by use of printed circuit boards are shown in the background of this montage. (*Courtesy Owens-Corning Fiberglas Corp.*)

finish.   Part B shows workers inserting woven insulating glass fiber material between the cores of a large electrical motor. Molded fiber glass polyester slot wedges are also used in similar applications.

General uses of fiber glass in reinforced plastics have been fairly thoroughly discussed in the previous segment. In most of the uses in RP/C in which cloth or fabric is laminated together with other types of reinforcing intermediates, the cloth is usually employed as a facing or containing vehicle.   However, when used singly (an all-fabric laminate) highest strengths result because high glass contents up to

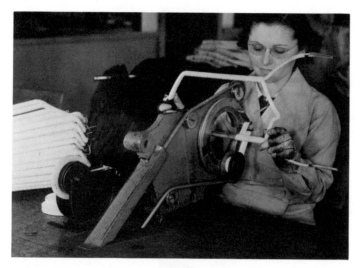

(A)

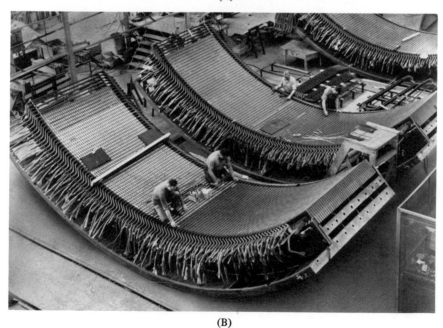

(B)

Fig. 5-28. Part A. Here an operator is winding fiber glass tape around an assembly of leads for an electric motor or generator. Part B. Workers inserting woven fiber glass insulating material between cores in assembly of a large electric motor. RP/C molded slot wedges are used in conjunction with the varnished woven insulation in this application. (*Courtesy Owens-Corning Fiberglas Corp.*)

65–70% are reached. RP/C structures such as radomes and other aircraft parts require 100% fabric laminates. The stability of all-fabric laminates is also desirable in tooling for checking fixtures, master tools, and other types.

In the combined reinforcements for RP/C applications, fabric is used in conjunction with chopped strand mats, woven roving, and veil mat. These include boats, sporting goods, chemical tanks, ducts and piping, and in fabrication of exotic aerospace components.

(A)

Fig. 5-29. Part A. New vaulting records were established in the sports world immediately with the advent of the highly flexible but rugged fiber glass vaulting pole. This photo shows the pole in maximum flexure about to project the young vaulter up to as "close to heaven" as he'll probably get for quite a while—or hopefully a little closer on the ensuing jump. Part B. Several types of orthopedic casts made from knitted fiber glass fabrics. (*Courtesy Owens-Corning Fiberglas Corp.*)

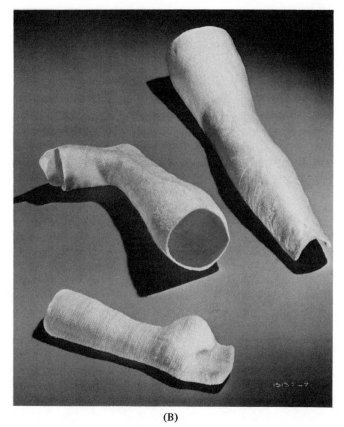

(B)

Fig. 5-29. (*continued*)

Newest applications include portable tennis court surfaces and skateboards.

Frequently, fabrics or bands of fabrics are impregnated with resin, then wrapped on a tapered mandrel and cured to form such parts as hollow fishing rods or the vaulting pole pictured in Fig. 5-29, Part A.

Orthopedic Casts

Great beneficial use has been made of glass fabrics in orthopedic casts and prosthetic appliances. Orthopedics deals with correction of deformities of the skeletal system. Prosthetics concerns fabrication of artificial parts for therapeutic treatment of such deformities or physiological shortcomings and deficiencies.

Due to their greater softness, original flexibility, and porosity, knitted fiber glass fabrics are frequently used in orthopedic casts. Examples are shown in Fig. 5-29 Part B. Prosthetic appliances are usually fabricated using fiber glass fabrics with a polyester veil mat for surfacing inside and out.

Reinforcing for Paper, Films and Foils

Scrim fabric, as explained, is made by coating warp or lengthwise yarns with a hot-melt adhesive (usually asphaltic) and applying the crosswise strands prior to the time the hot-melt adhesive solidifies. The scrim process provides an inexpensive, nonwoven, fabric-type material with good tear strength in all directions, plus excellent resistance to moisture, fungus, and rot. Hence films, foils, and papers with scrim fabricated between layers maintain their shape and resist dimensional change. Many of the coverings or membranes for fiber glass insulating ducts are fabricated using scrim laminated in (see Pipings and Ducts). Figure 5-30 shows use of scrim in reinforcing and binding two plies of kraft paper.

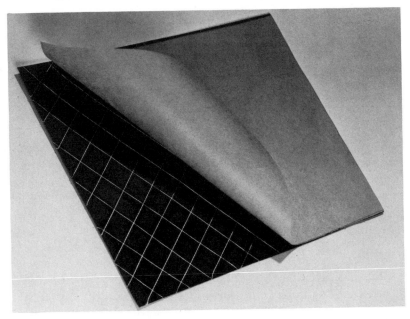

Fig. 5-30. Use of scrim (nonwoven) fiber glass fabric in laminating together two plies of kraft paper. Scrim is also used to join together and reinforce films and foils, or mixtures of the three materials. (*Courtesy Owens-Corning Fiberglas Corp.*)

Coarse Industrial Fabrics

Glass fabrics are used as a strengthening and bonding agent in abrasive wheels. Usually laminated flatwise in the plane of the disc or circular cross section of the wheel, the abrasive media and bonding glue are placed in between the fabric plies. By virtue of using the fiber glass fabric, the abrasive wheel is made much more safe and functionally effective.

With the increased limitations being placed on use of asbestos, coarse fiber glass fabrics are superseding that material in many areas. One notable instance is use as gasketing in microwave ovens. Also, as explained in Insulation for Transportation Modes (Marine), glass fabrics are used as lagging cloth to cover navy hullboard employed to line bulkheads, cabin liners, etc., in marine construction.

One recent innovation which is purported to grow to major market status is use of coarse glass fabrics as a carpet and floor tile backing material. Use of woven roofing fabrics is also in the ascendency (see Fig. 5-31, Part A).

Fabrics for Decorative and Clothing Applications

The readers' attention should again be called to the decorative fiber glass fabrics used in automotive Topliners,® and on acoustical office partitions and dividers (see Insulation in Transportation Modes (Automotive) and Acoustical Insulations, respectively). The 401® or "bouclé" weave is used to provide varied and interesting surface textures in these fabrics.

A detail of use of the sliver type of woven fabric for wall decoration is shown in Fig. 5-31 Part B, and photo of an actual installation is shown in Fig. 5-32.

In Fig. 5-33 is shown a use of lightweight fiber glass fabric made into drapes for homes, hotels, or offices to provide privacy yet admit light. Such fabrics have found almost unlimited use because of their interest and appeal to tastes of both decorators and users, their ease of handling, freedom from cumbersome maintenance, and qualities of stability and fireproofness. Such fabrics are "coronized" or heat-treated to relieve the stress of the weave, thereby preventing abrasion between contacting fibers and greatly lengthening the life of the product.

In Fig. 5-34 Parts A and B is further illustrated the unlimited fabric possibilities created by varying weave and dying techniques.

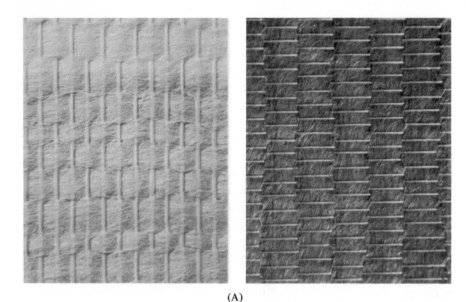

(A)

(B)

Fig. 5-31. Part A. Use of sliver in weaving a fabric suitable for a roofing media. Part B. Close-up of woven staple fiber (sliver) prepared for use as wall-covering material. (*Courtesy Schuller GmbH, Wertheim, West Germany, Subsidiary Johns-Manville Sales Corp.*)

Fig. 5-32. Use of woven staple fiber (sliver) as an effective wall decoration on the wall panels on which the mirrors are mounted. (*Courtesy Schuller GmbH, Wertheim, West Germany, Subsidiary Johns-Manville Sales Corp.*)

Fig. 5-33. An installation of fiber glass draperies in a hotel hallway, emphasizing their highly desirable properties of excellent appearance, permanence, ease of maintenance, stability, and fire resistance. *(Courtesy Owens-Corning Fiberglas Corp.)*

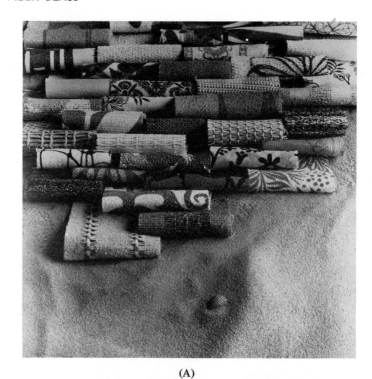

(A)

(B)

Fig. 5-34. Parts A and B. New fiber glass fabric designs made possible by varying weave and yarn dying. Part C. Aluminized Beta® fiber fabric used as protective, fire-resistant clothing. (*Courtesy Owens-Corning Fiberglas Corp.*)

(C)

Fig. 5-34. (*continued*)

The utility or practical side of use of fiber glass fabrics for protective clothing has been made possible by exploitation of the full technical capabilities of beta fiber and yarns. Originally widely proposed for bedspreads, curtains, draperies, and some items of personal clothing, the functional uses which have persisted are exemplified in Fig. 5-34 Part C, showing clothing to protect against heat and fire.

Use of Fiber Glass in Screening

Fiber glass twisted and/or plied yarns may be coated with polyvinyl chloride plastisol or organisol by dipping and die-wiping, and curing in a long oven through which the strands are traversed by winding them onto the finished packages outside the exit end of the oven.

Originally developed for identifying electrical tracer yarn, an infinite variety of strong and pleasing colors may be produced. However, most screening yarn is colored neutral grey, as would be expected.

Conversion of this vinyl-coated material into a plain-weave fabric, approximately $\frac{1}{16}$ in. centers warp and fill, has provided a screening material highly competitive to all other conventional materials. Very recently, fiber glass 'comfort" screening has been developed in which vinyl-coated strands wider and flatwise oriented (woven perpendicular to the plane of the screen) perform to admit the same amount of light, but to block or "shade" excess sunlight, thereby providing savings in air conditioning energy and greater comfort to the in-dwellers.

In the very early days of promotion and promulgation of fiber glass for many of its uses which are now widely accepted on a matter-of-fact basis, vinyl-coated screening was one of the products placed under hard scrutiny and evaluation.

Samples were woven and sent to the military, who were keenly interested in its promise and likelihood of improved properties. The technical committees of the then major fiber glass producing companies studied the results of the lengthy and exhaustive army tests with more vigor and discussion.

With fiber glass screening, gone were the difficulties in cleaning and great tendency of dirt to stick to the screening material. Gone was the paint discoloration due to corrosion of metal and subsequent run-down of wash onto painted woodwork, siding, and sills usually found with metal screening. Gone also were several other undesirables.

Only one adverse test result was returned. When the fiber glass screening was stretched as a cover over a bait box full of crickets, their strong mandibles were able to make holes in the screen due to the friability of the glass fiber when flexurally stressed, and also due to the crickets' ability to negotiate through the thin vinyl coating.

In one committee meeting, this report induced great chagrin, until one technical wag rebutted and rejuvenated all flagging aspirations with the statement "Oh—that's perfectly all right, we're going to put the screening in a non-cricket area!" He must have been correct, because to the best knowledge available, crickets have not interfered with or slowed the acceptance of vinyl-coated fiber glass screening material.

The material is used not only for conventional domestic and commercial window screening where and when necessary, but many other applications exemplified by the following: enclosures for home, commercial, and apartment swimming pools, patios, doors, and porches. Industrially, it is gaining in favor, and is used in weather-

Scratch Test

Take a quarter and run it across the face of a piece of non-Fiberglas screening. The result is a permanent scratch, a scratch that could have happened during packaging, shipping or installation of your screen. Now run the quarter across the Fiberglas screen. There are no marks.

Puncture Test

Stick a pencil through the mesh of a non-Fiberglas screening then try and return the screening to its original shape. You cannot. The hole will always be there. But when you remove the pencil from a piece of Fiberglas screening there is no hole. Just the same mesh pattern you started with.

Bending Test

Try bending a piece of non-Fiberglas screening material. It stays bent. And nothing you do will remove that crease. Fiberglas screening bounces back into its original shape.

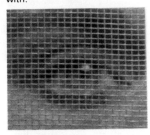

Visibility Test

Hold a piece of non-Fiberglas screening to one eye and a piece of Fiberglas screening to the other eye. You can see the difference in see-through visibility. There are other things you can do with Fiberglas screening, too. Kick it, punch it, drop a weight on it and it will out-perform any other screening material.

Fig. 5-35. Easily conductable tests which readily point up the advantages of vinyl-coated fiber glass screening over competitive types. (*Courtesy Owens-Corning Fiberglas Corp.*)

protective and seal-anchoring coatings for fiber glass insulated piping, tanks, vessels, and other equipment.[56]

In Fig. 5-35 are presented a series of simple tests and observations which readily point up the superiority and advantages of fiber glass screening over competitive types.

In Fig. 5-36 there is recorded for posterity a method of usage of fiber glass screening devised by your avid author for aiding and abetting the efforts of avid gardeners everywhere.

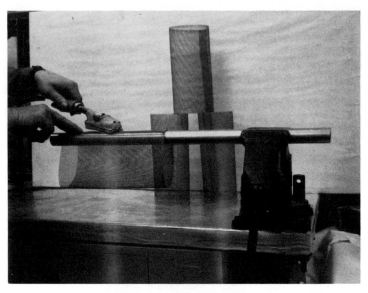

(A)

Fig. 5-36. Part A. Use of vinyl fiber glass screening as an assist to horticulturists. Portions of flat screening material approximately 13 X 8 in. are heat-sealed using a hot iron to form screen cylinders 4 in. in diameter (smaller if desired). Part B. Plant clippings, transplants, starts, or other live plant elements, in need of further development of their root stock prior to permanent planting, may be inserted into the cylinders and packed firmly with moist earth. They should then be submerged or surrounded with additional dirt in either a greenhouse or outside in the ground in a partially shaded or protected area. Many plant-containing cylinders may be planted in the temporary bed in close proximity. After half a season, or over winter, when the root stock has had ample time and opportunity to grow and strengthen, the entire cylinder containing plant and roots may be readily dug up and easily transplanted to a permanent site where necessary or desired. The fiber glass cylinder is simply left in the ground after transplanting. It does not restrict plant growth, because roots extend outward through the screen mesh, and the non-plant fraction ultimately disintegrates.

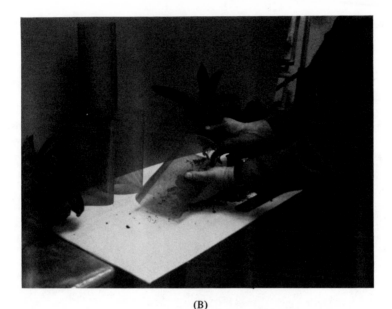

(B)

Fig. 5-36. (*continued*)

Filtration

The great advantages of maintaining clean air are becoming more and more like necessities, with pressure of Federal regulations and, actually, the ultimate goal of human survival. Baghouses that work like a giant vacuum cleaner have long been in existence. The bags are usually 3 to 4 ft in diameter and up to 40 ft long (fiber glass bags are fabricated by sewing).

In usage, an inlet manifold from the industrial operation, the air from which is to be cleaned, feeds the contaminated air to the open ends of the bags. The dust builds up on the inside of the bags, and efficiency of collection increases as the cake thickens. Periodically, for cleanout, the air flow is reversed, and the dust falls free and is collected for removal at the bottom of the manifold. Figure 5-37 illustrates an array of fiber glass fabric bags in a large conventional baghouse.

Many types of fabrics other than fiber glass are used to fabricate bags for dust collection. These include cotton, wool, nylon, propylene, acrylic, Nomex,® and Teflon.® Fiber glass possesses the best

Fig. 5-37. An array of fiber glass filter bags 3 X 40 ft in a conventional bag-house. (*Courtesy Owens-Corning Fiberglas Corp.*)

temperature resistance properties, dimensional stability, lowest elongation (not over 4%, hence no bulging), and may be more easily heat-set.[57]   Particles from submicron to 10 μ diameter are readily and reliably removed using fiber glass filter bags, and the efficiency averages 99.9%.   Many other advantages of fiber glass accrue, the most pertinent of which is the fact that less corrosion of ancillary metal parts results because the fiber glass bags can operate above dew point (temperature at which condensation will occur in the gas stream).

Other applications of fiber glass fabric in filtration are in nuclear containment, in which cloth-wrapped fiber glass insulation is used as a seal and retainer in nuclear steam piping.[58]

In the realm of filtration of molten metals to remove slag and impurities in the casting process, fiber glass fabrics have long enjoyed successful application in the processing of molten aluminum metal.

High Temperature Fabrics

Complementing the discussion of High Temperature Insulating Fibers, continuous-filament textile fibers may be produced from refractory ceramic melts.[59] Compositions include eutectic melts of combined (1) oxides of aluminum, boron, and silicon; (2) oxides of aluminum, silicon, and chromium; and (3) oxides of zirconium and silicon. The compositions and individual properties are not yet fully released, but maximum temperature resistance extends to 2600°F, and their moisture absorption factor is 0.06% maximum.

Main uses developed to date are the superseding of metal belts in oven processing of electronic components, and in thermocouple wire insulation (see Fig. 5-38 Parts A and B, respectively).

(A)

Fig. 5-38. Part A. An oven belt woven from continuous-filament refractory ceramic fibers used to supersede metal in throughput-oven processing of printed circuit boards. Part B. Continuous-filament refractory ceramic fibers used as braided-sleeve wrapping or insulation around thermocouple wire. Maximum continuous temperature resistance is 2600°F, with short term resistance to 3000°F. (*Courtesy 3M Company.*)

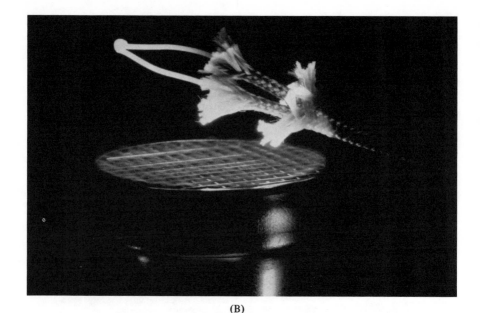

(B)

Fig. 5-38. (*continued*)

Other previously developed uses of glass fibers for high temperature resistance include ironing board covers, furnace and oven liners or insulation protection, and like applications. Standard E-glass continuous-filament fibers possess temperature resistance up to 1200°F.

Fabric Structures

Although contemplated for centuries, the potential use of flexible membranes as elements of construction has only been seriously experimented with for the past 30 years. Even more recently, the combination of the plastic Teflon® impregnated into fiber glass fabric has set in motion a completely new and innovative concept of construction. This flexible coated fabric material can be emplaced as the roof of a building, stadium, swimming pool, or as an entire inflated structure at one-fifth the cost of current conventional building methods.

Domes for several large public stadiums, college student activities centers and fieldhouses, large exposition buildings or pavilions, community centers, and many others have been recently conceived, planned, and executed with great success. See Fig. 5-39, Parts A, B, and C.

In fabric structures, elevation of the flexible membrane may be maintained either by forced air pressure, created and sustained within the space enclosed, or by a network of regularly positioned tension members (cables, etc.) to serve as supporting means, and over which the coated fabric is draped.

Use of Teflon®-coated fiber glass in these air or cable supported structures provides many advantages, the most important of which are: (1) incombustibility and non-support of combustion; (2) resistance to weather, and a projected 20 year life; (3) translucency; (4) overdesigned strength to resist vagaries of the weather, including high windstorms and snow loads (coated fabric tensile strengths up to 2000 lb/in. of width are possible); (5) ease of fabrication and maintenance; and (6) low year-around operating costs. The Buck Rogers'

(A)

Fig. 5-39. Part A. Civic and athletic stadium, Pontiac, Michigan, covered by Teflon®-impregnated fiber glass becomes one of the original fabric structures of our modern era, and its success presages wide application in the future. Capacity is 80,000 spectators, and area covered is 10 acres. The fabric structure is supported by force of blower air from within. Part B. A view from within the stadium looking upward at the coated fabric. Part C. The unusual design of the fabric-structured Student Activities Center at La Verne College near Los Angeles, California. (*Courtesy Owens-Corning Fiberglas Corp.*)

(B)

Fig. 5-39. (*continued*)

(C)

Fig. 5-39. (*continued*)

prognostications in the 1920's of covered cities and protective enclosures of other large unit areas may be a reality in our lifetimes!

Other examples of flexible fabrics enjoying relatively wide usage are depicted in Fig. 5-40. Illustrated are flexible fabric awnings for homes (Part A), and utility tarpaulins (Part B). Many other uses are possible in any area where it is desirable to lengthen fabric life or service by preventing abrasion. Such uses as attention-getting display flags or streamers, and wide horizontal or vertical awnings or louvered blinds are highly practical and enjoy wide acceptance.

YARN TYPES AND APPLICATIONS

A wide variety of fiber glass yarns are produced, hence an elaborate, detailed system has been worked out for their identification. This precise terminology is explained in appropriate manufacturers' literature,[60] and classifies the following in a combination letter-numerical designation: (1) type glass composition (E, C, or S); (2) type of yarn (continuous-filament or staple); (3) filament diameter (letter designation corresponding to actual filament size in microns

(A)

(B)

Fig. 5-40. Part A. Impregnated fiber glass fabric used to provide a flexible awning for a residence. Part B. Impregnated flexible fiber glass fabric used as a tarpaulin for protection of farm materials temporarily stored outside. (*Courtesy Owens-Corning Fiberglas Corp.*)

or inches); (4) strand weight of bare glass (expressed in strand count which equals yards per pound ÷ 100); (5) relationship of twisted strands to plied strands (expressed 2/2, 3/0, 4/12, etc).

For finished continuous-filament yarns which comprise stranded fiber glass textile products, following are the approximate ranges or limits of variation in the most essential parameters: (1) strand count—1800 to 15 (180,000 to 1500 yd/lb), (2) filament designation—letters B, C, D, E, G, and K (twisted and plied strands made from filaments coarser than K-diameter are subject to excessive filament breakage in the twist and ply operations); (3) number of filaments per strand—varies from 51 to over 2000.

An untwisted, unplied accumulation of strands drawn in parallel orientation into a package weighing 10, 20, 35 or more pounds is termed roving. A varying number of basic strands may be accumulated, providing the finished roving strand a yield which ranges between 750 yd/lb (G-fiber, 20 ends) and 125 yd/lb (K-fiber, 60 ends). Other combinations are possible, including roving produced directly from the bushing (usually 225 to 625 yd/lb). Binder type is usually designated in roving products, while filament diameter is not.

The term "staple fibers" designates that a bulk or loose sliver strand has been twisted or twisted and plied. Staple fibers are produced using C, D, E, G, and J-fibers and the same yarn nomenclature applies.

The bare glass yards per pound of fabricated (twisted and plied) yarn may be approximated by dividing the actual yards per pound (strand count $\times$ 100) by the total number of strands (twisted plus plied strands). Yards per pound are reduced slightly both by twisting and plying operations and by the addition of sizes and binders.

In fiber glass manufacturing, textile packages are wound onto a wide selection of supply packages for later use in the applications to be explained. These include: (1) twister tubes (elliptical build); (2) ply tubes (with milk-bottle build); (3) straight and pineapple cones; (4) spools; (5) bobbins: (6) serving and braider packages, some with ends tapered, or streamlined; (7) beams; and several others. The usage of these will become obvious in the discussion of applications. In Fig. 5-41 is illustrated a group of representative fiber glass yarns and typical packages. Also included are some characteristic end products.

Fig. 5-41. Types of textile yarns and packages on which they are purveyed. Some finished laminated products are also shown. (*Courtesy Owens-Corning Fiberglas Corp.*)

The unusual and unique properties of glass fibers as a textile material are clearly evident in the descriptions of the product usages, elucidated in the following segments. Subjects discussed are both the general and specific uses for yarn, and include cordage and cabling, polymer- and wax-bonded strands, sewing thread, dyed yarns, yarns for paper reinforcement, pressure-sensitive tape, textured yarns, mixed yarns, carpet backing, electrically conductive yarns, metal-coated fibers, uses for staple fiber, fiber glass yarn abrasives, use as ground cover, cement and gypsum-board reinforcement, and impregnation of and incorporation into rubber and other elastomers. Specific yarn types themselves as well as major and unusual applications are considered.

Yarns, Cordage and Cabling

The major use of fiber glass yarns is in fabrics, discussed in the prior section. The second major use is in electrical applications. Filler components, tension members, and braided or wrapped coatings

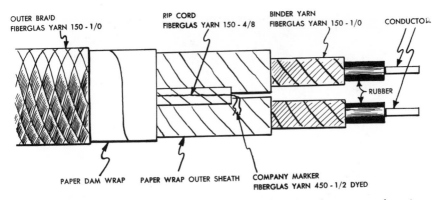

OUTER BRAID
FIBERGLAS YARN 150 - 1/0

RIP CORD
FIBERGLAS YARN 150 - 4/8

BINDER YARN
FIBERGLAS YARN 150 - 1/0

CONDUCTOR

RUBBER

PAPER DAM WRAP   PAPER WRAP OUTER SHEATH   COMPANY MARKER
FIBERGLAS YARN 450 - 1/2 DYED

Fig. 5-42. Diagrammatic cutaway illustration showing the ways various types and configurations of fiber glass yarns are used in electrical cable. (*Courtesy Owens-Corning Fiberglas Corp.*)

constitute the highest volume usage. Figure 5-42 shows a diagrammatic cutaway view of use of fiber glass yarns in a sheathed building cable, made essentially nonmetallic by virtue of use of vitreous fiber material. The heavier yarn (150 4/8) comprises cordage used as the tension members. Cordage may be sold plain or with a neoprene coating applied.

Tying cord, and many other uses requiring high strength and/or abrasion and fraying resistance are natural applications for fiber glass cordage. Knot-holding ability makes the material useful for many critical tasks.

Another example of the abrasion resistance of fiber glass yarns is in its use as an asphalt-spreading mop in built-up roofing and repair. See Fig. 5-43.

Wax and Polymer Bonded Yarn and Strands

Wax-impregnated yarn on serving and braiding packages is used as an outer braid for wrapping electrical wire and cable. The wax treatment lubricates and minimizes abrasion during the braiding operation (see Fig. 5-25).

Wax-bonded strand is made using a multiplicity of yarn ends impregnated with melted microcrystalline wax and die-wiped to establish overall diameters ranging between 0.024 and 0.070 in.

Wax-bonded strand is also used as a filler in electrical cabling as depicted in Fig. 5-42. Waxed strands are also consolidated (com-

Fib. 5-43. A fiber glass mop for spreading hot-steeped saturating asphalt in built-up roofing and roof repair applications. (*Courtesy Owens-Corning Fiberglas Corp.*)

pacted) and coated with a highly flexible plastic sheathing for use in drapery traverse cords and also high tensile strength clothes line.

A cellular plastic coating is applied to surround fiber glass strands and make them suitable for weaving indoor upholstery fabrics.

Also included in this category are the PVC-coated yarns used to weave fiber glass screening (see Fabrics).

Sewing Thread

Fine yarns and Beta® fiber are used in critical sewing applications. Notable properties are individual fine-yarn flexibility coupled with tensile breaking strengths up to 31 lb, excellent knot strength, and high temperature resistance. Notable uses are in sewing all types of filtration equipment including microfiber filter elements, and also the large fiber glass fabric air filter bags shown in Fig. 5-37. Use of fiber glass thread in sewing insulation is shown in Fig. 5-44.

Dyed Yarns

Using one of the earliest fiber glass treating processes developed, yarns may be dyed in a large spectrum of strong, solid colors. Actually, the starch sizing and not the fiber glass per se, is dyed.

Fig. 5-44. Use of fiber glass sewing thread to encase aircraft ducting with quilted and faced insulation (also made of fiber glass). (*Courtesy Owens-Corning Fiberglas Corp.*)

Originally forming the base for colored decorative curtains, drapes, etc. (now silk-screened or otherwise colored), use of the dyed yarns has persisted. Current major application is in providing identification in electrical cables (see Fig. 5-42).

Other important applications for colored or dyed yarns have been more recently established. One end of glass yarn colored bright, intense red is used as a tracer strand in spray-up roving. In this case, the colored material is a polyester-compatible coating, and not merely starch. When chopped into short lengths (1 to $1\frac{1}{4}$ in.) and projected onto the mold together with the roving and simultaneously delivered polyester resin, the tracer strand provides for the operator a reliable gage of thickness and distribution of the glass reinforcement. Red and blue tracer strands are also used in chopped-strand mat reinforcing products for identification of and distinguishability between very close mat weights (e.g., 1 and $1\frac{1}{2}$ oz).

In answer to the question, "why not color the glass composition to be fiberized?" undoubtedly raised by many persons, it may be stated that a commercial venture to do just that is written into the record books. In the 1930's, the Libbey-Owens-Ford Glass Company (now Libbey-Owens-Ford Company) developed and marketed a product known as "Vitrolux" in which colored fiber glass strands were sandwiched between two panes or "lights" of glass in the manner in which safety glass is produced. The intent was to provide a translucent building material similar to the present-day RP/C architectural paneling (Fig. 5-9). Production costs were high, the material was heavy, and versatility was lacking. Therefore, the product "came a cropper" soon after its inception and introduction.

The glass fibers used were drawn from colored rods. Regardless of the color intensity in the base glass, the color in the fibers was hazy and pastel in shade due to external light being reflected from the fiber surfaces. Other deterrents to color being physically incorporated within the fiber body are quality control (uniformity and consistency of color), melting difficulties, and also the fact that a large melting facility would be tied up, being relegated to one color only.

Paper Reinforcing

In addition to the scrim fabric for paper reinforcing discussed under Fabrics, single-end packages of fiber glass yarns are creeled and drawn into a laminate composed of top and bottom layers of strong kraft paper. Asphaltic, hot-melt plastic or other bonding medium is added. The strands are laid down in longitudinal, cross-direction, and diagonal orientation. Water-soluble gum or glue is applied to the bottom surface, and the laminate is split into 1, 2, 3, or 4 in. wide tapes for use in automatic dispensers.

Single-strand, nontwisted roving is also laminated between light outer plies of cardboard to strengthen construction and permit lesser amounts of paper to be used in packaging such goods as beer in six-packs, etc.

Fiber glass reinforcing stabilizes and strengthens the paper and prevents elongation. A high-volume market exists for gummed, reinforced paper tape products, and many other uses are gaining acceptance rapidly.

Pressure-Sensitive Tape

Twisted yarn only is beamed and the strands fed parallelwise in a process designed for production of pressure-sensitive tape.

In the technology of beaming, twisted or plied strands are taken up under constant, equal tension onto a large-diameter hub with rigidly affixed disclike sides. Lengths of 2-, 3-, or 4000 yd of each strand are applied in parallel orientation. Beams containing plied yarn are used for weaving fabrics.

A strong plastic film backing is used as the carrier. The fiber glass strands are impregnated with mastic as they meet the carrier film and are pressure-applied by doctor blade since hot-melt application would distort the carrier film.

Strand density is approximately 64 strands/in. Tapes are made in widths of $\frac{1}{2}$, $\frac{3}{4}$, 1, and 2 in. In Fig. 5-45 the great strength of fiber glass pressure-sensitive tape is shown by its support of a small, early-model Corvette sports car. The nominal tensile breaking strength of each strand (150 1/0 yarn) is 5 lb. Supporting the car are eight bands of tape, each 2 in. wide, hence containing 1024 total strands. At 5 lb tensile strength, total resistance is 5120 lb which makes the eight bands of the tape amply over-designed to support the car. The curb weight of this model Corvette is 2850 lb.

Textured Yarns

A system was devised for intermittent drafting of one or more (but not all) components of a plied yarn to introduce small incremental looped segments which are longer than the base strand. The drafting is accomplished by air jets. These short incremental lengths become

Fig. 5-45. The inherent strength of fiber glass yarn is vividly pointed up in this illustration of a car being supported by eight bands of 2 in. wide reinforced pressure-sensitive tape. (*Courtesy Owens-Corning Fiberglas Corp.*)

looped and stand out from the longitudinal axis of the strand. The result is a "nubby" strand termed bouclé, or Taslon.®

When woven, these strands produce a tridimensional, bulky fabric with pleasing decorative effects. Inorganic dying pigments may also be incorporated so that the fabric may be coronized (heat set) without altering the coloration (see Fig. 5-34).

Mixed Yarns

To provide the benefits to the fiber glass of increased abrasion resistance and permanent protection by fusion, Dacron® polyester yarns are twisted together with the fiber glass.

The resultant mixed-yarn strand is wrapped onto magnet wire and cable jacketing to provide electrical insulation. When subjected to mild heating, the polyester yarn softens, fusing the fiber glass to the wire base in a fairly permanent composite. The mixed yarn is also used to form housings for small electrical batteries, with the polyester yarn subsequently fused to bond the fiber glass into a permanent structure.

Yarn for Carpet Backing

Augmenting the use of fiber glass sliver and staple fiber for carpet backing (as discussed under Fabrics), continuous-filament yarns are heavily impregnated and encapsulated with an elastomeric polymer to make them suitable for this application. Figure 5-46 presents a composite view showing the yarn on a large package, the woven fiber glass carpet backing, and the top pile or tuft side of the completed woven carpet.

The fiber glass material contributes several major improvements in weaving conditions and finished product properties briefly discussed as follows: (1) The irritation in handling sometimes associated with the use of fiber glass is drastically minimized, and "fly" is eliminated, with no loose ends or strands becoming separated as the yarn enters the loom; (2) the coated strand provides the carpet with dimensional stability, resisting shrinking and stretching, and eliminating humps caused in other backings by inadequate adhesion and high humidity; (3) the high strength of fiber glass permits expanded use of carpeting into areas previously regarded as too severe for nonglass carpet

Fig. 5-46. Composite photo showing a package of yarn for carpet backing (heavy elastomeric coating is readily noticable), the bottom side of the finished woven carpet (upper right) and the top pile or tuft side (lower left). (*Courtesy Owens-Corning Fiberglas Corp.*)

backing; (4) probability and occurrence of rot, mildew, and bacteria growth in the carpet-backing yarn is greatly reduced, since the coated fiber glass material will not absorb moisture; (5) friction created by the coating develops the highest interlocking tuft-binding strengths, equal or superior to natural and synthetic organic-fiber carpet-backing yarns.

Electrically-Conductive Yarn

The advantages of dimensional stability, low elongation under tension, and good electrical insulating properties (no arcing, etc.) render fiber glass extremely useful in this application. A proprietory conductive coating is applied, cured or set, and the fiber is covered with rubber sleeving to form a cable.

The main usage is in the cabling from distributor to spark plugs, and the market saturation is almost 100% in the automotive field. Approximately 6 ft of conductive cable are used in almost every car on the road.

The benefits result from the controlled resistance of the conductive cable. The resistance may be varied to satisfy several requirement levels. Static from the distributor and other outside influences is virtually eliminated, and this results in suppression of noise so that in-car AM, FM, and citizens-band radios operate more clearly and efficiently.

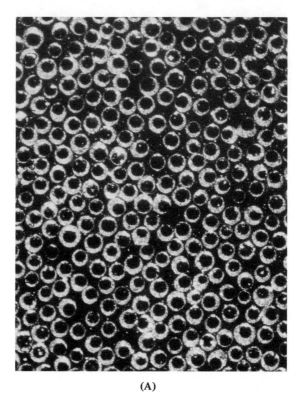

(A)

Fig. 5-47 Part A. An end-view photomicrograph at 150X of aluminum metal coating on glass fibers. The coated fibers are nominally 0.001 in. in diameter. Part B. The aluminum-coated fibers for chaff aligned on a carrier film being spooled into a package for shipment. Used in providing false targets for radar, these fibers are packed into a bomb which may be shot from land, ship, or a plane, and is set to explode at the desired altitude, releasing the chaff (see also Fig. 5-48). *(Courtesy Tracor, Inc., Austin, Tex.)*

Metal-Coated Fibers

As described in Chapter 4, fibers may be directly coated with molten metal (usually aluminum) immediately after fiber forming from the hot-melt bushing. Metal coatings of aluminum and others may also be applied as a secondary operation (coating under vacuum) to fiber glass yarns and fabrics from which the original protective sizing has been removed by heat-cleaning.

The metal-coated fibers are approximately 0.001 in. diameter. They have two major uses: (1) military; and (2) laminated into plastics for electrical shielding and thermal conductivity.[61]

The military uses comprise airborne or dispersed chaff to confuse and deceive enemy radar and optically guided missile systems. Radar chaff, in the form of tuned dipoles, is an inexpensive countermeasure to combat radar-guided weapons systems. See Fig. 5-47.

(B)

Fig. 5-47. (continued)

When laminated into either thermoset or thermoplastic resins, using any of the known molding methods, and using the identical RP/C process tooling, metallized fibers alter the usual laminate dielectric and thermal properties and induce conductivity.

Electrical shielding is provided in the following areas: electromagnetic interference (EMI), electromagnetic compatibility (EMC), and electrostatic discharge (ESD). Benefits are also gained by being able to control or design into an RP/C laminate different levels of thermal conductivity.

Although somewhat more expensive than uncoated reinforcing fibers, laminated strengths are equivalent, and no loss results in mechanical properties.

As regards applications in the area of improvement of electrostatic properties, laminates containing metal-coated fibers are used in critical housings and structural parts of data-processing equipment. Electrical conductivity in the treated-fiber laminate is in the range 0.1×10^5 ohm cm., and conductivity may be specifically controlled by adjusting fiber length, concentration, and other processing variables.

Use of conductive fiber in RP/C eliminates the secondary operations necessary on other equipment such as sheet-metal liners, plating, coatings, or conductive paints. Some potential also exists for use of these materials to prevent hazardous electrostatic sparking in hospital operating rooms and other similar areas.

When RP/C is rendered thermally conductive by laminated-in metal-coated glass fibers, several benefits accrue. The cure cycle of the laminate and its subsequent cool-down will be shortened by the improved conductivity. In finishing applications, these laminates are made receptive to electrostatic paint-spraying techniques and other electrostatic finishing. Also, when in service, rapid thermal transfer away from locally heated areas makes these composites less susceptible to cigarette burns and other thermal scars. See Fig. 5-48.

Uses for Staple Yarn

Supplemental to the usage in woven products (see Fabrics), sliver products have many utilitarian applications. Bulk fibers, unsized,

Fig. 5-48. This photo shows a hank of metal-coated fiber as removed from the spinning or winding drum. When chopped and laminated into thermoplastic or thermoset polymeric materials, these fibers induce both electrical and thermal conductivity by virtue of a multiple number of fiber-to-fiber contact points along the filament length and in the three-dimensional laminate network. Applications for plastic laminates incorporating these metal fibers are in EMI shielding in the fabrication of cases for computers, peripherals, instruments, small electric motor cases, CB radios, telephones, etc.; for electrostatic grounding of all types of devices including equipment cases, clothing, rugs, conveyor belting, flooring, etc.; low-resistance current-carrying devices in electronics, and heating elements or tapes. Due to the inherent thermal conductivity, laminates reinforced with metal-coated fibers may be more rapidly heated and cooled, thereby improving processing factors and production rates in injection and compression molding; thermally conductive laminates improve heat transfer of instrument cases and thus reduce requirements for forced-draft circulation. (*Courtesy MBAssociates, San Ramon, Calif.*)

are used for air and liquid filtration, and for sterile pharmaceutical wadding and packing material. Liquid filtration includes use in aquariums. When sized, the sliver intermediates go into fabrics for the uses described and also for thermal and electrical insulation such as for turbine generator blankets.

Other uses of staple fiber strands include fillers in electrical cabling (see Fig. 5-42) and a wrapping for thermal insulation and protection around automotive hoses and tubing. See Fig. 5-49. Thermal and acoustical insulation in automotive and industrial mufflers is achieved by using sliver as an internal wrapping or packing.

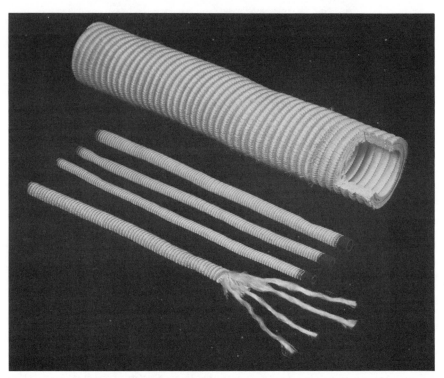

Fig. 5-49. A variety of sizes of glass fiber tubing prepared on circular weaving equipment using no-twist staple fiber yarn as the core. These tubular parts are used in automotive applications as thermal insulation around metal gas-line piping for temperature control in the automatic choking system. (*Courtesy Frank D. Saylor and Son, Inc., Birmingham, Mich.*)

Abrasives from Continuous-Filament Yarn

Finite-sized glass fibers (0.002 to 0.004 in. diameters) are grouped together in parallel bundles, coated with an elastomeric material to reduce sidewall rubbing and deterioration, and used as abrasives by rubbing the exposed fractious glass ends against the desired surface.

Uses include the following: (1) erasing pencil or ink impressions from paper, carbon copies, and photocopies (dispenser-type housings provided); (2) for corrections on metal offset plates, cleaning at the same time for improved printing; (3) to make corrections on all types of fluid duplicating masters in office work and printing; (4) spot cleaning on metals, such as in pattern, tool and die work, and other industrial applications; (5) burnishing and cleaning printed circuit boards, artware, art and industrial ceramics, precious metals, jewelry (see Fig. 5-50).

This application, although falling into the "small users of large samples" category, is a lucid example of the versatility of fiber glass, and presages the many interesting and greater uses yet to come.

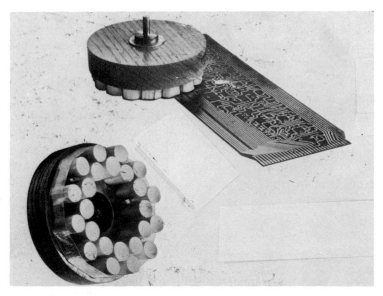

Fig. 5-50. Illustration of the use of bundles of continuous glass fibers as an abrasive for cleaning printed circuit boards. (*Courtesy Rush Eraser Co., Inc., Syracuse, N.Y.*)

Ground Cover

Several uses have been developed for fiber glass continuous-strand roving apart from reinforced plastics applications (RP/C). One of the most interesting and beneficial of these is use as a ground cover to stabilize the soil and prevent erosion.

This activity is carried on using large-scale outdoor equipment that dry-sprays the strands drawn simultaneously from 10 to 12 roving packages. After a road or bank area to be treated is conventionally graded, shaped, seeded, and fertilized, the mat of sprayed continuous glass fiber roving is applied in a random configuration using an air-powered applicator gun mounted on a field truck. Following, a tack-coating of asphalt or the equivalent is applied to loosely bind the roving strands. The result is a porous mat that holds the soil in place, preventing erosion while the grass seed is taking time to germinate and grow. The application of fiber glass roving to a graded area, and subsequent coating with an asphalt tackifier are illustrated in Fig. 5-51.

Even following the full maturation of the grass stand, the roving continues to prevent erosion by remaining in situ, hence providing a ground additive that is environmentally acceptable. Reportedly, this ground cover system is less expensive and requires less labor than sodding, or applying jute mesh or excelsior blankets. Also, experience has shown that it will not harm other plant or animal life indigenous to the area.

Many applications have been made to date, and include athletic fields, golf courses, highway banks and berms, airports, and industrial sites. Many other areas would benefit from the fiber glass roving ground cover treatment, including construction sites, county and municipal land or building projects, land reclamation, and graded ski and recreational sites.

Fig. 5-51. Part A. Application by spray of fiber glass continuous-strand roving to a roadway berm and portion of a hill following grading and planting of grass seed. Part B. Addition of asphalt tackifier to provide a loose, random, porous mat through which the grass may grow. The glass fiber ground-cover coating continues to stabilize the soil and prevent erosion even after the grass has fully grown, thereby being totally acceptable according to environmental considerations and standards. (*Courtesy Owens-Corning Fiberglas Corp.*)

(A)

(B)

Reinforcement for Gypsum Board and Concrete

Three major categories are involved in coverage of this subject: (1) reinforcement for gypsum board; (2) glass-reinforced mortar for coating and erecting cement block and similar walls; (3) incorporation of fiber glass reinforcing directly into concrete structures.

Gypsum Board

Chopped fibers of E-glass are incorporated into the water slurry of calcined gypsum which is then spread between chip paper cover sheets and supported while the gypsum hydrates and hardens. The fiber glass is reduced to monofilaments and disperses readily, providing good uniformity of coverage in the mixed batch.

The benefits to gypsum board contributed by the fiber glass reinforcing are: (1) an improved reinforcement over other types (asbestos, organic fiber); (2) improved impact strength; (3) improved fire resistance, since the glass fibers are also incombustible, and resist breakage of the gypsum board when subjected to thermal shock (being rapidly heated from one side); (4) there is no drop in board strength after long term aging of the gypsum board in continuous service. Hence, it may be readily noted that addition of fiber glass greatly improves the performance of gypsum building board.

Glass-Reinforced Mortar

A system of surface bonding of masonry walls was devised in which reinforcing fiber glass is added to a mortar mix, and the mix applied directly by either trowelling or spraying to a tightly laid-up new concrete block wall (both sides) or to an existing wall for resurfacing.

The mortar is a mixture of the usual water-setting cementitious materials in addition to the reinforcing fibers. In this case, the glass fibers must be of an alkali-resistant composition to withstand the rigorous chemical environment induced by the concrete mix (see Table 5-1).

In carefully controlled building-panel tests conducted by the National Concrete Masonry Association Research Laboratory, surface-bonded masonry wall structures were determined to be completely satisfactory as a building element. Using the alkali-resistant glass on both sides of the wall in coatings $\frac{1}{8}$ in. thick (3.2mm) it was found that the flexural strength was increased by a factor of 2 over a conventional mortar-joint wall, and the impact strength was 20% greater. Although the compressive strength was 65% of that of the conventional wall, it was still within allowable design loads for most concrete block construction. Also, the surface-bonded wall can be completed in only 58% of the time required for a conventional mortar-bonded masonry wall.[62]

Surface-bonded masonry walls have been evaluated and found to possess the following combination of properties: (1) adequate strength, and resistance to thermal shock over several winter periods; (2) water resistance (provided by an added chemical); (3) four-hour fire resistance (8 in. block); (4) sound insulation, providing more acoustical protection than fully mortared block walls; (5) major model code approved by city and state agencies; (6) a wall unit equally acceptable in comparison with tilt-up walls, poured-in-place walls, or metal buildings; (7) various pleasing surface finishes applicable after either trowelling or spraying on.

This mortarless reinforced surface finish has been applied successfully and is widely accepted for numerous constructions including old and new buildings, in mines, repair of structures, and others. In Fig. 5-52 is shown use of erection of a mortarless concrete block

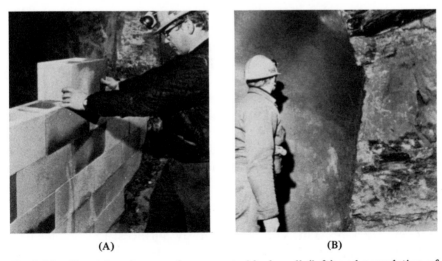

(A)   (B)

Fig. 5-52. Mortarless lay-up of a concrete block wall (left) and completion of the AR glass reinforced surface-bonding layer (right) to complete an air-stoppage wall in a mine. (*Courtesy Owens-Corning Fiberglas Corp.*)

wall and completion of the reinforced surface-bonding layer as a vitally necessary air-stoppage system in a mine.

Glass-Reinforced Cement (GRC)

Since 1960, considerable activity has been extant in fiber-reinforced concrete. Primarily the major concern was use of steel fibers as the reinforcement. Several researches were conducted using E-composition fiber glass as the reinforcement, but its lack of required chemical durability against concrete (or cement) has been known since 1951. In the period 1966 to 1970, alkali-resistant fiber glass compositions together with cement and concrete reinforcing systems were developed in the U.K. by Pilkington Bros. Ltd. and B.R.E. Today 50 or more licensees have been accumulated, and tests on the glass-reinforced structures are being carried out worldwide to determine long-term durability, strengths, effects of weathering in many different climates, and other parameters. Currently OCF and Ferro Corporation hold licenses granted in the U.S.A.

Manufacturing methods are based upon processing premixed glass fiber and cement (casting, extruding, injection molding, etc.), or by the simultaneous placement (spray-up) of chopped fiber and cement mix. The random planar arrangement of fibers in the spray process apparently provides more effective use of reinforcement than the premix technology. In the spray process, the deposited mass is subsequently dewatered and deaired by suction (evacuation by vacuum), providing superior mechanical properties. Hence the spray evacuation method is well suited for production of sheet stock which may be either cured flat or further processed by molding or contouring while still in the green state. Glass loading is usually in the 5% range, but may be varied up to 40%.

Figure 5-53 shows heavily profiled, $\frac{3}{8}$ in. thick, AR glass-reinforced cement panels being assembled as decorative architectural panels on a new building.

In addition to the spray cast-and-evacuate panels, following are other examples of use of GRC: (1) highway overlay, to provide thinner, crack-resistant concrete highway surfaces; (2) replacement of steel mesh reinforcement in general utility items, such as beach houses, architectural building panels, sewer-line liners, septic tanks, burial vaults, parking bumpers, drain pipe, roofing tiles, downspout

Fig. 5-53. Alkaline-resistant fiber glass is used to reinforce these contoured, diamond-shape, $\frac{3}{8}$ in. thick cement panels used as a decorative architectural paneling on a car-parking facility in Stourbridge, U.K. (*Courtesy Cem-Fil Corporation, Nashville, Tenn.*)

splash blocks, transformer-TV towers, air conditioner pads, meter boxes, water troughs, picnic tables, artificial rocks, and decorative fascia, and in railroad ties, (steel rebar rods are retained where originally used).

Properties of GRC may be summarized as follows:[63] (1) Vapor permeability—2 perms or less; (2) moisture permeability—no moisture transfer through a sheet from rain driven by 73 mph wind; (3) moisture and expansion—0.14% or less at 25% sand; (4) high and low temperatures—no adverse effects for 25 cycles of 12 hr at 68°F to 12 hr at 14°F (frost cracking test for asbestos-cement structures); (5) thermal expansion ranges from 7.2 to 11.1 × 10⁻⁶/°F between 77 and 115°F; (6) sound insulation—for a 0.375 in. skin of GRC, 4 lb/sq ft, sound reduction index = 22 dB at 350 Hz and 39 dB at 4000 Hz; (7) fire rating—fire rating depends upon matrix type and method of manufacture, but the optimum conditions produce fire rating of over 2 hr.

Reinforcing of Rubber and Other Elastomers

Use of continuous-filament fiber glass to reinforce rubber vee-belts constituted the first foray into this market. Other successful applications are as treated cord for carcass, bias belting or radial belting in tires, and as treated chopped strands for reinforcement of tire treads, solid tires, and as elastomeric functional and decorative fascia.

All fiber glass products for reinforcement of rubber and elastomers are treated with the same resorcinol formaldehyde latex applied to other types of tire cord. The general properties and improvements imparted to the rubbers by incorporation of fiber glass are: (1) increase in both flexural and tensile (stiffness) moduli; (2) greater dimensional stability; (3) better resistance to wear and abuse; and (4) less heat buildup in actual performance. Additional advantages will be pointed out in discussions of each individual sphere of application.

The initial patent for fiber glass reinforcement of rubber products was issued in 1939,[64] and covered use in bias ply, bias belted, and radial belted tires. The treatment used on fiber glass for reinforcement of elastomers is similar to that applied for the carpet-backing material described earlier in this section.

Fiber glass cord for rubber reinforcing comprises twisted yarn only. In the heaviest yarns supplied (2 ends of ECG-75's, 5/0 together), the yards per pound equals 621, and the average breaking strength is 103.0 lb direct pull. As received, the RFL content will be approximately 15 to 17%.

Vee-Belts, Timing Belts, and Associated Applications

Fiber glass loading in vee-belts is usually on one plane close to the outside or back of the belt. Excellent resistance to static loading, elongation-inducing stress, and heat is accomplished, as illustrated in Fig. 5-54. The belt containing organic fiber became considerably distended when subjected to a laboratory "torture" test in which the fiber glass-containing belt remained unaffected. However, experience has shown that fiber glass-reinforced vee-belts do not withstand severe shock-loading of the type induced by sudden starts of automotive and industrial equipment. Although the low percent elongation of glass fiber (4%) prevents distortion of the belt, this

Fig. 5-54. Results of a laboratory "torture" test showing distortion and elongation resulting in a vee-belt made using synthetic organic fibers. The belt made using fiber glass maintained its original dimensions. (*Courtesy Owens-Corning Fiberglas Corp.*)

property is the cause of broken filaments and strands when the tensile strength is momentarily exceeded.

Hence, a more successful application has been in timing belts. These are apparently not shock-loaded to the same degree. A greater fiber density is possible, and the belt also contains neoprene teeth, a belt backing, and a nylon facing, in addition to the reinforced rubber carcass.

Other homologous applications are in (1) FHP vee-belts; (2) poly vee-belts (a multiplicity of grooves affixed to one belt body or backing); (3) reinforcing strands braided around steam hose and externally coated; (4) in continuous belts for snowmobile tracks.

Tire Cord

There is considerable merchandising, publicity, and advertising pressure associated with automobile tires and the comparative performances of the various built-in reinforcing materials: rayon, nylon, polyester, fiber glass, and steel wire. A lifetime study would be required to accumulate sufficient data to speak with even semi-authority; and this would be complicated by the frequent, rapid

changes in direction taken by the tire market. Fiber glass has as-
sumed an enviable stature as a material of tire construction, and its
properties and performances will be described.

A fabric is formed as the tire-cord reinforcement, and is heavily
impregnated with additional tire-compatible, rubber-bonding material. The RFL-treated fiber glass performs well in the looms and is
readily adaptable to conventional tire-cord fabric-weaving techniques. Also, it is fully compatible with either natural or SB rubber.
In Fig. 5-55 is illustrated a cutaway section of a fiber glass-reinforced
tire, showing bias belts beneath the tire tread.

The first breakthrough for use of fiber glass in tires were the items
carrying the Polyglas® label—an American-made tire consisting of a
polyester fiber carcass and fiber glass bias belting under the tread.
Fiber glass bias belting continued in favor and, along the way, fiber
glass material for the carcass was also added.

The first fiber glass radial belts were placed into tires by Armstrong
in the U.S. about 1966 and by Kleber Columbes of France in about

Fig. 5-55. Cutaway view showing bias-ply belted fiber glass tire reinforcing
material beneath the tread. Radial-belted fiber glass is coming on strong and
provides a slightly smoother ride than the bias-belted tires. (*Courtesy Owens-
Corning Fiberglas Corp.*)

1969 and were used on winning cars in prestigious European races in 1972 and 1973. Since then, bias-belted fiber glass tires have grown in their threat to steel-belted tires, and radial-belted fiber glass tires are growing in usage and popularity.[65]

Properties of tires and tire performance by fiber glass plies and belts are summarized in the following: (1) a softer ride, with radial belts representing a slight improvement over bias belts; (2) lower reinforcement cost—$1.97 per tire for all fiber glass vs $3.12 for steel-polyester; (3) readily available without danger of short supply, as compared to petroleum-based synthetics nylon, rayon, polyester, and Kevlar, and as compared to steel wire, which is in equally short supply; (4) in addition to the previously mentioned improvements of increased moduli, hardness, and less heat buildup, a tread backed up with biased or radial fiber glass exhibits better traction.

Fiber glass tire cord also exhibits equal or superior performance in long-distance driving tests: (1) cut-resistance after 23,000 miles (cut through tread and part of the fiber glass plies, glass withstood the test—steel rusted); (2) low-inflation to 16 psi and driven for 10,000 miles, and other torture tests; (3) in a high-speed laboratory test running tires against a revolving cylinder, radial-belted fiber glass tires showed substantially less "squirming," or distortion of the tread than did synthetic-fiber reinforced tires, hence indicating longer running and less tread wear at higher speeds. All this translates into better performance, smoother rides, more resistance to damage and stability when striking chuck-holes, rocks, and road debris, etc., and greater resistance to heat buildup for the fiber glass tires.

An interesting corollary to use of fiber glass in tires is the development of an RP/C "run flat" device, intended to deduct additional weight from a car (only 12 lb per wheel added vs elimination of 60 to 80 lb for spare tire, clamping fixtures, jack tire irons, etc., plus the benefit of creating additional trunk space), and at the same time provide safety against blowouts.[66]

RP/C-molded circular segments with a C-shaped cross section are mounted on any standard wheel rim, and extend to within 2 in. of the inside of the tire body when the tire is properly inflated. A hard surface of approximately 5 in. face width is provided to run on if the tire becomes flat. Speeds of 50 mph and travel up to 50 miles are reported to be possible without damage to the tire or tread. This device is currently experimental and will probably continue for some time in the development and testing stage.

Reinforcement of Rubber in Elastomeric Bodies and Thick Casts

Having its origin again with tires (treads, not carcasses) this usage generated itself as a military "brat" or upstart. Now it enjoys wide usage in not only tire treads, but many solid-cast rubber products which call for resistance to deformation and good wearing qualities.

The fiber glass used is in the form of chopped strands $\frac{1}{4}$, $\frac{1}{2}$, or 1 in. long. They are also treated with resorcinal-formaldehyde latex, or resorcinol-formaldehyde-neoprene latex, and may have a silicone chemical added to promote better adhesion. As for rubber for automotive tire rubber, individual filaments are thoroughly impregnated with the coating material, and maximum adhesion is gained with either natural or SB rubber.

Why is this development important, and how did it really get started? Have patience, friends, and together we'll unravel an unusual bit of interesting fiber glass history.

In flying large C-130 cargo aircraft into Vietnam in 1966, during the height of the conflict, much damage occurred to the then soft tire treads upon landing on shrapnel-pocked strips and jungle runways, encountering foreign objects and debris, and even gunfire damage. Needed was a greatly improved tread, i.e., firmer, harder, and more cut- and penetration-resistant. Incidentally, all aircraft tires are designed for multiple retreading, so that a better tread would have been highly beneficial and adaptable.

The suppliers of the fiber glass tire cord available at that time were requested to enter into a joint development program with the aircraft tire manufacturer. The first suggestion was to use the chopped strand form of reinforcement, incorporating it into the tread compound at approximately 20% by weight.

Initial testing showed much-improved penetration resistance, increased tread life by a factor of 1.3, better blowout protection, and greater general durability. Although their use in Vietnam was short-lived, and not much viable field data was accumulated, the tire manufacturers knew they had created a new "giant." Tires for the mammoth American SST were subsequently planned, but the SST program was cancelled.

However, the technology spilled over into commercial aircraft, since problems in runway configuration and tire design were not well coordinated. Because "wedges" of water built up on flat runways when wet, aircraft landing or taking off at high speed would virtually "plane," becoming hard to control and losing braking power. Longitudinal grooves cast into the concrete runways solved the "planing" problems, but chewed up ordinary soft-tread tires, inflicting a defect classified as "chevron cutting." Aircraft tire treads are circumferentially slotted, and do not possess a tread such as is incorporated into automotive tires. The larger planes, DC-10 and 747, were affected the most.

When rubber-impregnated chopped fiber glass strands were incorporated into the tread, somewhat oriented, the circumferential grooves would not "close in" under landing pressure as would occur in the soft tire tread. Also the tread was harder all around, hence an increased resistance to any type of skidding also resulted together with greater protection to the tire casing. Silane crosslinking agents were also added here to the RFL coating to promote even better adhesion between the glass and rubber. The beneficial result is that this type of reinforcement is now being used in treads of tires for almost all commercial planes flying regular runs, i.e., MacDonnell-Douglas DC-8's, DC-9's, Boeing 707's, 727's, 737's in addition to the DC-10's and 747's on which it was first tried for commercial flight improvement. Tires for fighter aircraft are also being retreaded using the fiber glass-reinforced rubber treads.

Use in aircraft tires is only the beginning, however. Applications are indicated for a host of related products, including solid rubber wheels and tires such as for industrial service vehicles, in piping, gaskets, extruded rubber products, load-bearing bushings, shock absorbers, rubber household appliances and utensils, belting which would be favored by cross-directional reinforcement, conveyor belt surfacing, and many other areas.

One additional benefit for use of fiber glass reinforcing in hydrocarbon rubber compounds has been the development of Nordel® EPDM (ethylene-propylene-diene-monomer) thermoset elastomeric resins. These possess highly desirable and heretofore not available qualities such as low cost, resistance to deterioration by ozone, oxygen, weathering, heat, and chemicals, and others. These elastomers also possess high tear and tensile strengths, plus excellent dynamic and low temperature properties. When reinforced, a large-percentage improvement results in all basic properties, notably toughness, recovery from deformation, impact and distortion resis-

Fig. 5-56. Use of glass fiber reinforced EPDM rubber in this 1977 Oldsmobile Starfire sports car is in protective and functional parts that also present exceptionally good appearance. The molded front-end parts which are car-width immediately above and below the bumper as well as the black elastomeric protective strip across the bumper itself are glass-reinforced EPDM. The molded fascia parts surround the headlamps and turnsignal lights and are also formed into a grill for admission of the cooling air. They accept automotive paint and blend well with the other car parts. Being elastomeric, they yield to hand pressure, but are strong enough to withstand low-speed impact. (*Courtesy E.I. du Pont de Nemours & Co.*)

tance.   The EPDM material also is compatible with automotive paint.

Its major use to date has been in decorative, protective, and functional fascia, such as that depicted in Fig. 5-56 including use in automotive bumpers.   Prognostications are for 500 million pounds of EPDM to be used annually in a few years, 20% of which would consist of fiber glass chopped strand rubber reinforcement.

FIBER OPTICS

Introduction

There are three general major areas of usage of glass fibers as light conductors.   These all require filaments that are drawn into the essentially continuous configuration.   The uses fall into the categories of: (1) decorative; (2) light and image transmission; and (3) communications.

Although it was first shown in 1858 that a still-standing transparent fluid medium (water in this case) would transmit light along a curvilinear path, it was not until 1927 that basic patents were issued recognizing the concept that images could be conducted through fibers and bundles of fibers.   Some attempts were made in 1930 by German researchers to produce an image-transmitting fiber bundle, but in 1950, the problems of isolation of the actual light-carrying fiber were recognized, attacked, and solved.   It was found necessary to retain the light and/or image within the boundaries of a specific fiber by coating it with another medium of lower index of refraction (again, glass is best).   Since this innovation, important commercial development of fiber optics has proceeded with rapidity.

These three areas of usage of fiber optics are briefly described here.   Technical reasons for their performances are studied, and representative examples are presented and explained.

Decorative

An unusual method of using continuous glass fiber has been devised in order to furnish a decorative luminary.   For use in home or office, these items are intended to fill a design focus or interest spot, or to be used as an attention-getter, eye-attractor, conversation piece, or like application.   They are not necessarily intended to provide a quantity of light for reading.

Individual fibers approximately 0.006 in. thick are drawn onto large drums 5 to 6 ft diameter.   They are then cut to desired length, assembled in bundles, both ends polished, permanently bound at the base, and mounted vertically in a housing and over an adequate light source,[67] and permitted to hang free and unbound at the top.

The fibers comprising the outer peripheral portion of the bundle, being unsupported, become influenced by gravity, and bend in flexure forming a graceful arc.   The fibers farther toward the inside of the bundle drape or fall to a lesser degree than those at the outside of the bundle.   The fibers at the center core of the bundle stand almost erect without any draping at all.

Light from the source beneath the fiber bundle is transmitted upward and outward through each individual fiber, yielding a distinct and pleasing glow at each fiber end. The result is that the glowing fiber ends combine to outline the shape of a solid geometric figure, hemisphere, or paraboloid, as defined in space by the glow of the fiber ends.

Originally, the rounded "head" or "bouquet" of the fibers was tediously created by handcutting the fibers in the bundle into different lengths, with the outer fibers being left longer so that they would drape or bend in flexure to a greater degree. Recent technological improvements have been made[68] in which a heating operation is added during manufacture of the spray. Heat induces softening of the fibers in the bundle, causing controlled bending.

The fibers in the outer portion of the bundle, being exposed to the full temperature of the heater, bend to a maximum arc, while those successively located toward the center of the bundle bend to a lesser degree. The heating is closely controlled, again leaving the center fibers essentially unaffected by the heat treatment. There is naturally a lesser amount of residual strain due to bending in a bundle of fibers which are heated than in a bundle which are bent only due to the force of gravity. In Fig. 5-57 is illustrated a representative Luxury-Lamp® fabricated from the bent fibers, mounted in a housing and lighted from beneath, as described. Decorative fibers to be subjected to the bending procedure described are made with filament diameters slightly thicker up to 0.025 in. This is done for purposes of greater mechanical durability and increased light-carrying capacity.

Many other variations may be built into these decorative fiber optic units. The overall height of the table models varies from 11 to 24 in. Floor-standing models are also made 36 in. high with a clear acrylic plastic dome to protect yet show the fiber spray. The fiber spray may be made to revolve slowly, adding interest and an eye-attracting facet. Various solid colors may be applied, or a rotating varicolored element may be incorporated to successively change the hue projected upward through the fibers. Also much variation may be built into the base to provide compatibility with many furniture styles and to suit widely differing personal preferences.

Illumination is being improved, with the incorporation of high-intensity quartz-halogen light sources. These units possess the advantages of intensity of illumination and interest equivalent to or better than those for lighted candles, but of course without the fire hazard accompanying an open flame.

Fig. 5-57. Photo of a Luxury Lamp® fabricated from bent fibers lighted from a center core or gathered bundle of fibers. Light of various or even mixed color shades may be used, and the shape or form outlined in space by the fiber ends may be widely varied to provide most interesting decorative effects. (*Courtesy Fantasia Products, Division of Valtec Corp., West Boylston, Mass.*)

A fiber optic lamp of this type placed on the top of a television set, and lighted during set operation and viewing times, would assist in preventing maximim pupillary dilation in the human eye. This is desirable and helpful in avoiding eye fatigue and damage because the light-to-dark intensity variation on a television screen is greater than that on a moving picture viewing screen by a factor of approximately 10.   Hence, it is advantageous to have an additional light source facing the eye while viewing TV to prevent the maximum dilation that would occur in a near-darkened or side-lighted room.

Fiber optic designs or decorative displays have been incorporated also into furniture-module housing to form interesting memorabilia such as antique cars or bicycles, home bar or rec-room signs, auxiliary desk or night-lights, etc.

Obviously no great scientific challenge was required to devise and purvey these decorative fiber optic elements. In fact, the external light-retaining low-index fiber sheath required for image production and communications is not necessary in these decorative applications. However, it is perfectly obvious that these interesting adaptations of fiber glass will bring pleasure and interest to purchasers and, over the long haul, be economically rewarding for the developers and manufacturers.

Light and Image Transmission

The isolation or insulation of individual fibers or filaments in fiber optics alluded to above was found necessary for maximum light retention within.   Efficient conduction of light through fibers is made possible by total or near-total internal reflection from the inner walls.   Even though smooth, the surface "spills" light to the outside due to minor surface imperfections, defects, grease, dirt, etc.

The most efficient method of retaining fiber optic light within the fiber wall has been determined to be coating or cladding the fiber with a glass of a lower index of refraction.   Hence, to produce the fiber of double composition, the transmitting glass is melted as a rod inside with the insulating glass outside as a tight-fitting tube.   This assembly is placed under vacuum, and indexed slowly into the heating area of a vertical furnace where it melts and seals, permitting the composite coated fiber to be attenuated at high speed.   The difference in index of refraction is determined by end-use needs (see Table 5-3).

With coated fibers, even though fibers touch in a bundle, no light is lost.   Reflection efficiency is approximately 99.9% compared to only 99.5% for metal-coated fibers.

TABLE 5-3.   Indices of refraction of combined core and coating glasses for fiber optic production and their corresponding end applications.

Material	Core index	Coating index	NA*	Applications
Glass	1.52	1.48	0.35	Faceplates, light guides for near U.V.
Glass	1.62	1.52	0.56	General purpose light guides, fiberscopes
Glass	1.66	1.52	0.67	CRT faceplates
Glass	1.81	1.48	1.04	Image tube faceplates
Plastic	1.49	1.37	0.58	Light guides
Plastic	1.59	1.49	0.56	Light guides

(Courtesy American Optical Co., Fiber Optics Div., Southbridge, Mass.)

*NA = numerical aperture

Fiber diameters from 0.001 to 0.004 in. are satisfactory for image or light transmission elements intended to withstand bending, such as in a cable. Heavier fibers (even rods, steel clad) are used for special nonbending fiber optic elements.

Individual fiber production, clad, is a viable and most effective technology. However, multiple-fiber drawing of up to 1000 clad preform elements has also been accomplished. Plastic cladding is satisfactory for some uses.

Plastic optic fibers are also produced. They have greater flexibility and are less brittle than glass, but have less efficient light-transmitting capacity. They also possess much lower resistance to heat, thus limiting their usage. Glass can be safely bent to a radius only 50X individual filament diameters and plastic fibers to a radius as little as 3.5X filament diameter. Hence, glass fibers of diameter 50 μ or less are required for small bending radii of bundles in a cable.

Light-gathering power of a fiber optic fiber or element, as in lens optics, is a function of the numerical aperture, N.A.[69] A meridional ray entering the end of the transmitting fiber must be incident to and reflected from the core-to-cladding interface at an angle greater than the critical angle for total reflection. Skew (off-center) rays incident at angles larger than those required to produce the critical angle must have a perfectly symmetrical circular cross section of the fiber to prevent dispersal. Aperture angles vary with glass composition between 24 and 128°.

Fiber optic elements (bundles of fibers) can be randomly arranged (not collimated) and form what is classified as an "incoherent" bundle. When precisely oriented at both ends of the bundle, the fibers will properly transmit images, and are termed a "coherent" bundle.

Incoherent fiber optic bundles are used only to direct light to necessary stations in instruments, control panels, and electronic displays in a greatly expanded array of equipment in many fields. These include medical, aircraft, computers and business machines, electronic gear including TV sets, and many other industrial, transportation, and commercial areas.

In coherent, image-transmitting fiber optics, the glass bundle is termed a "multifiber" bundle, and is comprised of a number of optically distinct 10 μ diameter fibers fused into a single strand. The multifiber elements may be combined into any desired pattern or band.   A multifiber bundle 25 mm in diameter contains about 3.5 million individual fiber optic elements, and is capable of resolving approximately 50 line pairs per millimeter.   Either noncoherent or coherent bundles may be sealed or fused only on the ends to permit bending when formed into a jacketed cable.

The largest area of application of fiber-optic image-returning elements is probably endoscopy in medicine.   Examinations and treatment of body cavities such as the esophagus, stomach, bronchi, and sigmoid area (large bowel) have been greatly enhanced and made infinitely simpler because of the adaptation of fiber optics.   Prior to their use, with a straight-line rigid gastroscope, only three-fifths of the stomach could be visually examined.   Using fiber optics, not only can the entire stomach be brought into view using the flexible probe, but the fiber optic head may be passed beyond the duodenum and reversed in direction so as to view abnormalities outside the stomach such as duodenal ulcers.   Figure 5-58 illustrates (Part A) a fiber optic gastroscope in use, and (Part B) a photograph of a characteristic image of a stomach ailment viewed using the scope.

A separate "sleeve" of fibers is used to transmit light to the object, to be reflected back through the coherent fiber bundle to a viewing eyepiece or camera lens.   Focusing means, wide-angle objective lenses, biopsy forceps, and many other beneficial ancillary appurtenances may be incorporated.   Large usage also is found in inspection of remote areas of industrial or engineering equipment.   Probes up to 15 ft long are used (see diagram of construction, Fig. 5-59).

Other uses in the medical and associated scientific fields include: (1) apparatus for eye surgery; (2) chemical analysis systems; (3) a probe for blood colorimetry analysis in a multichannel blood chemistry device; (4) in illumination components for new single-purpose analyzers such as the gastrological, gynecological, and proctological examination systems and assemblies, improvements have provided 16 times more brightness than is available with non-fiber-optic systems.

By combining clad elements and adding black glass coatings to each fiber to absorb and to eliminate stray light, optical face plates may be formed.   These are used as end windows in cathode-ray tubes for the purpose of recording and encoding directly on a film without the use of a lens.   Photometric efficiency is much higher than that for the lens system.   Bent-image conduits, tapered and fused, rigid bundles, and many other forms have been devised.

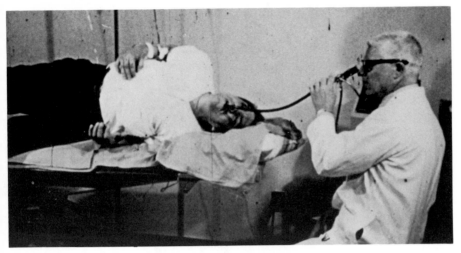

(A)

(B)

Fig. 5-58. Part A. A patient being examined with a fiber optic gastroscope. Part B. View of lower esophagus, as seen through a fiber optic esophagoscope, shows multiple superficial ulcers (light areas). (*Courtesy A. E. DaGradi, M.D. Chairman Gastroenterology, Veterans Administration Hospital, Long Beach, Calif.*)

Several other fields of application or marketing areas have benefitted from use of fiber optic illumination and/or image transmission in addition to those discussed. These include:

1. *Automotive and Transportation.* New Federal regulations specify that each functional switch or dial which provides operational or safety information to the driver or occupants must be illuminated.

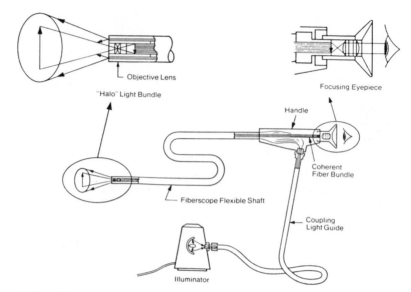

Fig. 5-59. Schematic diagram of construction of light-directing and image-transmitting fiber optic scope. (*Courtesy American Optical Corp.*)

Fiber optics, especially when woven into tape, provide the answer. Only one source of illumination is needed and it can be positioned in the best design and replacement location. Excess heat is eliminated, and less space than for many single lamp bulbs is required.

Traffic-control signals such as pedestrian STOP and GO signs, or aircraft runway and airfield signs and signals benefit from use of fiber optics. The light-conducting elements make possible increased brilliance of illumination, avoidance of deceptive phantom images, increased fail-safe reliability, and reduce problems associated with lamp replacement.

It has long been a goal of automotive engineers to establish a rear-view system for a car that would scan the entire field or panorama rather than a short, limited sighting through a window or externally mounted mirror. A video camera and television receiver would be one expensive solution. Another might be adaptation of a simple but effective fiber optic cable and enlarged, segmented viewing screen.

In air traffic control towers, movements of any plane on the ground or in the air once picked up are monitored so that the images are displayed to all controllers in the tower. This reduces the number of verbal exchanges necessary and helps eliminate possibility of errors. Fiber optics are also used to show runway clearances, approach paths, and for illumination of specifics on large visual display panels.

Also oriented to aeronautics, the image-transmitting capability of fiber optic elements has been well applied in a flight-training simulator. Miniaturization of runway lights through fiber optic cables made possible an operating display which matched actual takeoff and approach situations.

2. *Display*. In addition to the instances mentioned, alphameric characters or pictures may be posted in fiber optic cables or faceplated for display on notation boards. Unsafe hazards are eliminated since neither heat nor electricity is present. Road signs will pick up headlight glare and reflect it with greater brilliance than any presently used medium. Inside or indoor displays pick up room or skylighted light. Attempts are being made to channel sunlight indoors for illumination of displays.

3. *Instrumentation*. By means of fiber optic image transmission, the following uses in instrumentation have been made: (1) reading out-of-the-way gages or inspection of boiler tubes in engineering stations; (2) monitoring of temperature variation of turbine blades in aircraft engines through color comparison; (3) to detect flame failure in jet engines—the signalling source may be connected with an ignition mechanism which is triggered automatically by absence of light from the fiber optic element, thus avoiding the dangerous buildup of unburned fuel.

4. *Retail Sales Assist*. The extension of use of fiber optics to sales-code reading devices is a most interesting application. At the point of retail sales, an illuminating light pen is passed across the face of a printed line-band series affixed to each item. Not only is the value of the sale recorded, and the cost for several items totaled in a computer-register without button-punching, but at the same time the store inventory is automatically tabulated on a master sheet at a central location.

It is interesting to observe that almost all producers and packers of retail goods are applying the fiber optic code symbols on the original labels, although not all retail sales checkout points are fully set up with the automatic pricing and inventorying devices. Of course, Mr. and Ms. Consumer should always have the advantage of a printed final price label.

Fiber Optics in Communications

After Alexander Graham Bell invented the telephone in the 1870's, he also invented the photophone. Philanthropically or empathetically, he had persons who were hard-of-hearing in mind. This photophone transmitted human speech on a beam of light. The sun was the light source, and some fragile piece of equipment was jury-rigged as the modulator. Obviously, pursuit of this method of transmission or communication was wisely abandoned at the time by Bell in favor of copper wires and electrical transmission. For 100 years the entire world population has enjoyed and benefited from Bell's handiwork which resulted in the telephone.

However, in this modern era, the capability of the telephone and even our entire system of electrical communications is being overloaded because of under-capacity. Growth in the number of personal conversations and the conduct of business have burdened existing lines. Since initiation of "time share" procedures involving exchange of data between terminal points and computers, telephone lines have become even more overloaded. The results are delays in calls, interrupting of conversations, static and noise, all in spite of highly efficient systems like direct long-distance dialing, WATSBOX, etc. Another difficulty resides in the fact that, particularly in cities, space is at a premium and addition of extra communication lines is costly.

As pointed out in the preceding discussion of fiber optics in light and image transmission, the scientific world has long yearned to modulate electric power and microwavelengths to the higher-frequency light wavelengths because of their greater carrying capacity. This was made feasible only after development of the technology of cladding of optical fibers to prevent transmission loss.

Actually, information is sent either by analog modulation (continuous transmission) or by digital modulation (impulsively). In transmission on telephone lines, for instance, digital transmission is resorted to and only "bits" of the frequency line wave are modulated so that a higher volume of calls may be handled. These "bits" are put together on the receiving end and, because of the great number of signals pulsed, no gaps in the conversation are evident to the listener. Since light has a much larger wavelength band width than the lower electromagnetic waves, transmission by light has long been the objective, still using the pulse or digital modulation (see Fig. 5-60 Part A).

In the case of communications, many other possibilities were also considered. The question normally arises, "why not simply transmit through the air as for radio waves?" Actually laser beams, following their discovery, were considered a prime candidate. They could be sharply directed and might be modulated like radio waves. In practice, it was determined that the wavelengths of laser excitation were so short that dispersed atmospheric particles diffused the signals. The next step was to attempt to transmit laser beams through hollow tubing containing lenses to accomplish bending. This was abandoned because many lenses were required, making bending extremely limited and difficult.

As regards conditions which existed prior to lasers, only microwaves could reliably transmit a large number of bits per second in air. However, these were found too expensive, requiring frequent "booster" stations or towers, and they could not readily be protected against spying.

The discovery that closed the case for fiber optics as the prime candidate material to be used as a new direction for communications was first the 20 dB/km-loss cladded fiber rapidly followed by the 4 dB/km-loss fiber. The latter made possible transmission via fiber optics over direct cable for distances of 16 km without repeaters or signal boosters.

In addition to the larger waveband, the smaller diameter optical fibers occupy less cross-sectional volume than heavily-insulated copper wires. Also, complex electronic equipment for "chopping" into bits is eliminated.

So fiber optics, or the science of bending light, equally referred to as lightwave or optical waveguide communications, seems to have won an all-important first round as a material to solve existing problems and lead the revolution in communications. Many problems still exist, and coast-to-coast signal transmission systems are not yet in existence. However, very important applications have been made and new technology is developing at a rate which would put "gang busters" to shame.

In transmitting, an electric signal is passed through a preferred-type gallium-arsenide laser or a light-emitting diode (LED) and a current-drive attached directly to the fiber optic strand or bundle,

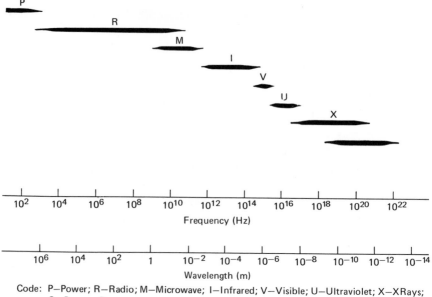

Code: P—Power; R—Radio; M—Microwave; I—Infrared; V—Visible; U—Ultraviolet; X—X Rays; G—Gamma Rays.

Conversion: $1\mu = 10^{-6}$ m; $1\mu m = 10^{-9}$ m; $1\text{Å} = 10^{-10}$ m; 1 mile = 5280 ft = 1609 m

(A)

Fig. 5-60. Part A.   Spectrum of electromagnetic radiations showing variations in frequency and wavelength for the several different measurable types, all of which have the same speed, C (approximately 186,000 miles/sec). This is not the entire range; in fact, some electromagnetic radiations with waves as long as 1.9×10^{7} miles have been measured. In fiber optics, energy from the "power" band is modulated up into the visible spectrum by exciting a gallium-arsenide or equivalent laser. This latter will vibrate almost monochromatically in the infrared or near-infrared (0.6 to 0.8 μ) and make possible transmission along fiber optic waveguides.   Part B.   A plot of signal loss in decibels per kilometer for substances of interest for comparision as fiber optic elements: W = pure water which is one of the most transparent in the visible range, but almost opaque in the infrared region; U = a bundle of glass fibers intended for transmitting light over short distances which is clearer than water over the longer wavelengths; FS = fused silica which serves satisfactorily in carrying light over long distances; PFS = pure fused silica which sets the Rayleigh scattering limit—almost perfect transmission in the infrared region. (*Part B From "Communication by Optical Fiber," J. S. Cook, copyright © 1973 by Scientific American, Inc., all rights reserved.*)

transmitted the desired distance as a light beam, reconverted to an electric signal through a photodiode, and fed directly to the end signal-handling circuit.   Pulses of as many as 44.7 million bits per second are being evaluated to determine optimum signal-handling feasibility of fiber optics.

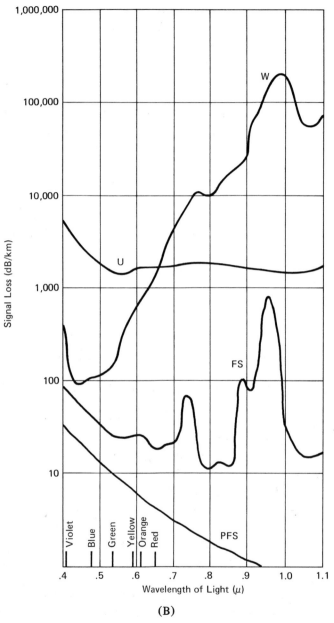

(B)

Fig. 5-60. (*continued*)

Much science and engineering have been responsible for the present well-advanced development of the signal input and output components. However, it is desirable here to consider the advances made in improvement of the light-signal-carrying glasses: (1) losses of the signal along a fiber are measurable as decibels attenuated per kilometer (dB/km), but a second system, the radiation transfer index (RTI), is also used. In a perfect cable or fiber (RTI = 1.0) all photons of light are transmitted, and in an absolutely imperfect cable, no photons of light get through (RTI = 0). Photons per second is a radiation flux that can be readily measured over an agreed-upon unit length. (2) problems of aberration or signal blurring, termed "differential delay," in which skewed or highly angular light requires longer time to traverse a single-clad optical fiber than do the collimated, coaxial meridional rays, were solved by development of multimode, or graded-index fibers as described in the captions for Fig. 4-9 Parts E and F (termed parabolic index or "selfoc"—self-focusing fibers) and also by maintaining core and cladding refractive indices as close as feasible; (3) problems of "material dispersion," in which light of one color travels through a fiber slower or faster than another (variable index of refraction for each color even though I.R. of the glass does not change) were solved by use of laser radiation, which is essentially monochromatic in the red and near-infrared range (lasers are also adaptable to focusing light into a small beam, or fiber end); (4) problems of high signal loss, dB attenuation, or attrition through a fiber were solved by learning how to melt exceptionally pure silica into optic fibers and "dope" the silica with selected chemicals to slightly alter its index (silica provides highest rates of transmission in the visible and infrared spectral regions as pointed out in Fig. 5-60, Part B); (5) many methods of joining or splicing fibers, including fusion, have been devised and minimal losses of 1 dB are usual; (6) problems of polishing fiber ends were solved by mechanically controlled fiber breaking in tension, yielding 100% mirror surfaces. Many other improvements and configurations are possible.[70]

Glass fiber optic waveguides possess many advantages when used as a medium for communications. The most salient of these are briefly described in the following: (1) the fibers are small and are fabricated into compact cables, hence they save installation space in crowded areas or cities where duct space is at a premium; (2) optical fibers are light in weight making them highly desirable for communications systems in aircraft, aerospace, missiles, and satellites; (3) no metals are involved, so that fiber optic cables are immune to static interference such as from electrical storms, magnetic fields, sunspots, nearby power cables etc., that currently have deleterious effects on wire and radio communications links; (4) the very limited dependence of optical loss on frequency and temperature in fibers greatly simplifies the design of receiver electronics, i.e., no equalization is required; (5) fiber optics are less costly than the present technology based upon transmission by wire, cable, and radio, hence are regarded to provide the same impetus to the communications field as transistors in the 1950's and integrated circuits in the 1960's; (6) signals may be branched off using optical couplers, and also multiplexing may be used, channeling several signals into one fiber or bundle of fibers; (7) repeaters or signal boosters would not be required every 50 mi. with fiber optics, as with wire or microwave communications systems; (8) as regards carrying capacity, in a $\frac{1}{4}$ in. cable containing 100 fibers, each fiber could carry simultaneously 1000 telephone calls, or several television programs; (9) signals sent over fiber optical waveguides are secure and the lines cannot be tapped, nor spied upon; (10) no shielding or grounding is necessary; (11) no hazards exist from sparks or fire, and the fibers will not attract lightning; (12) no failures or short circuits can be generated due to broken filaments.

Conversely, several disadvantages to the use of fiber optical waveguides are prevalent: (1) precise control of production parameters is needed to obtain near-ideal fiber dimensions and refractive index profiles (out-of-roundness increases skewness and higher dB or RTI losses); (2) although joinable, fiber or cable ends are connected with difficulty, and joints must be free from vibration, moisture, and dirt, which when present detract from signal transmission (also field splicing is difficult); (3) limited lifetimes exist for laser light sources and questionable reliability exists for some of the associated systems; (4) almost foolproof fiber protection is required to withstand the rough handling encountered in installation and maintenance; (5) the fiber optic systems should preferably be able to use fewer fibers and larger numerical apertures, plus lower losses, to permit longer transmission lengths without changing components; (6) problems are involved in getting light into the fiber cores due to the small diameters (only a few microns) of the cores.

As regards applications, many viable and problem-solving uses in communications have been made to date. Prognostications are for a $64.1 million fiber optic industry by 1980, with 54% in commercial communications, and $833 million in 1990 with 74% in commercial communications.

Noteworthy and direction-pointing uses of fiber optic waveguides in communication are represented by the following: (1) In telephone communications, an experimental 10.9 km (6.7 mi) repeaterless link was set up and satisfactorily placed in service with only 18 low-loss splices—no repeaters (boosters) were necessary. Intersystem transmission paths up to several miles long were included. The cable is only $\frac{1}{2}$ in. diameter, but can carry the equivalent of 50,000 telephone calls.[71] In the eyes of the Bell Labs evaluators the question is not whether the lightwave transmission systems will be introduced, but rather how and when they can best be implemented (see Fig. 5-61, Parts A, B, and C). An additional test or prototype fiber optic telephone system is set up in the city of Chicago; (2) an 800 ft fiber optic waveguide system was installed in a large metropolitan area (New York City) and transmits cable television signals from a roof antenna to head-end equipment 34 floors below; (3) communications in the entire post office system in the U.K. have been under study in the U.K. Post Office Research Center. By installation of an optical waveguide system it is hoped to eliminate the many inefficiencies encountered with the older wire-cable system; (4) regarding use as automotive components, a circuit network has been designed in the U.K for use of fiber optical electronic components to operate the complete ignition system; (5) in U.S. military department applications to date, the small size and light weight of fiber optic elements have been taken advantage of in installation of a sonar link on a submarine, and a central operating telephone station, both by the Navy. A total of 450 lb of copper wire was superseded using only 50 lb of fiber optical waveguide cable. Further applications in all the military branches include programs at least 5 years old, primarily data-bus applications mostly aboard aircraft, for military telecommunications, a fiber optic guided torpedo, and other applications geared generally to communications, weaponry, and surveillance;[72] (6) in Japan, optical fiber transmission systems for communication in and around power plants have been successfully evaluated and are being installed on a permanent basis in all major Japanese power plants throughout the late 1970's. The fiber optic

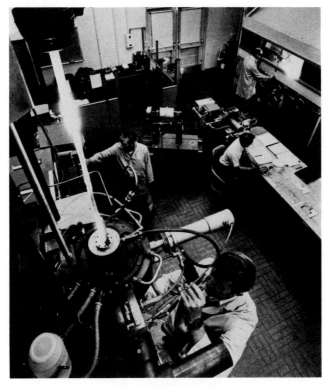

(A)

Fig. 5-61. Part A. High silica glass content optical fiber "preforms" are produced on the glass lathe (upper right) and then softened and attenuated into
thin fiber lightguides using the equipment shown in the foreground (see also
Fig. 4-9, Part E). Part B. Scientists and engineers check diameter of a fiber
lightguide cable emerging from the cabling equipment. The cable (reel of
completed material in foreground) is $\frac{1}{2}$ in. diameter, contains 144 fibers less
than 0.004 in. diameter and may carry up to 50,000 telephone calls. Part C. A
lightguide cable connector. When beyond the experimental stage, this metal
block between the fiber groups will be no larger than the cable itself, and will
align all 144 fibers in the cable to an accuracy of 1/10,000 in. (*Courtesy Bell
Laboratories, Inc.*)

systems supersede microwave radio communication, because the
latter requires heavily shielded coaxial cable that is not 100% succes-
ful in elimination of congestion in the radio frequency spectrum; [73]
(7) it has been determined that fused silica-core optical fibers can
transmit solar radiation energy for distances of approximately 40
miles. Possible applications are connection with and integration of

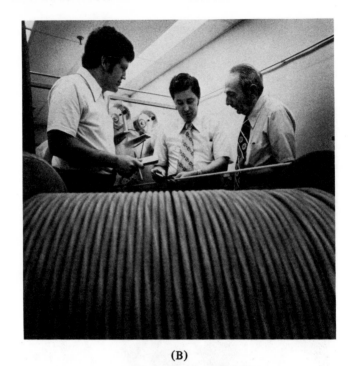

(B)

(C)

Fig. 5-61. (*continued*)

components for solar energy utilization, such as photo voltaic cells, solar-thermal power generation systems, plus solar heating, cooling, and illumination systems; (8) possibly reaching into the future of communications slightly farther than fiber optic waveguides will be integrated optics (also termed thin-film optics). These have the propensity, through differences in layered refractive indices, to trap lightwaves internally and become miniature electronic components with enormous information-handling capabilities.

Chapter 6 | *The Future of Fiber Glass*

INTRODUCTION

In order to appropriately culminate this text without simply stopping cold after the last bit of technical or applications data was recorded, comments from others were solicited. Knowledgeable persons who have spent all or a substantial portion of their careers in this field were invited to provide their opinions of the future of fiber glass. Those involved in both spheres of activity, insulations and reinforcements, were asked to comment.

These short pieces are included here together with identification of the contributor. These essays speak most eloquently in their own right for FIBER GLASS.

THE FUTURE FOR FRP COMPOSITES*

Change always comes about as a result of forces and conditions which apply economic and/or performance pressures on a material or a process. FRP/Composites is now about 35 years old and is a family of materials whose time has come. The very purpose or reason for FRP/Composites has often fallen on deaf ears, but now with the energy crunch and the basic need for improved performance we are moving into an area of maximum opportunity.

People buy a product because they either need it and/or want it. Today's opportunities go beyond a want situation—FRP/Composites is *needed*! Our basic job then is to satisfy these needs to the maximum benefit of all concerned.

*Prepared by Clare E. Bacon, Manager, Plastic Industry Relations, Owens-Corning Fiberglas Corporation.

314

The role of the designer or the design engineer has become more important and will continue to be critical in the proper use of FRP/ Composites.

Form always follows a material, and any new material is likely to go through a painful time of poor design. However, if we take a good look at FRP in use today the shapes we see support the dictum that a material's appropriate form can never become the material it replaces, but comes always from within itself.

Major advances and refinements in the development and production of fiber glass reinforced plastics (FRP) in the last 20 years have resulted in many new options for the designer. Design flexibility has always been the basic advantage inherent in FRP; however, in the early days feasibility factors such as strength, finishing, or high-volume production requirements often made it difficult to capitalize on the material's functional and aesthetic advantages. Fewer restrictions now exist as a result of improvements in mechanical, physical, and chemical properties, and in the economy and quality of production.

The engineer and the designer must take advantage of this improved set of conditions to pioneer new end uses for FRP/Composites. I am confident they will take advantage right now and move as rapidly as we as an industry give them the information and tools to help them accomplish their mission.

The automotive and transportation market for FRP has just moved ahead of the marine market as the largest user of FRP in the industry. This certainly gives us insight into what to expect in the next 25-30 years. FRP is no longer a volume-sensitive material replacing some metal on a temporary program, but is a material that is being considered first and is doing a better job than any other material ever used in that application.

And now for some predictions:

Some day you will see a total car body that protects the car and its occupants in the same way a football suit and pads protect a football player—not too rigid, not too flexible, just right to do the job. I think it will be a form of FRP when it happens.

Not too many years away are homes whose basic structures and components will be a form of FRP/Composites. Design will play a very critical role to minimize the negative threat of "all looking alike."

FRP/Composites in general will be fine-tuned for specific needs using a variety of reinforcements, fillers, resins, etc. The major R & D effort in the next few years will be concentrated in: (1) flammability; (2) new chemical systems to better satisfy government regulations; (3) more closed molding process development including lower-cost tooling; (4) weathering improvements; (5) improved process methods for better molded finishes; (6) FRP/Composite systems using an increased amount of inorganic components.

FUTURE USES OF FIBER GLASS INSULATION, ACOUSTICAL MATERIALS, ETC*

In the future, demand will increase markedly for fiber glass thermal insulation for solar collector panels, for solar heat storage tanks, and for fuel-cell power plants.

Due to continued and unstinting interest in health protection for industrial workers, especially in acoustics, improvements will be made in fiber glass acoustical panels for noise abatement, and their use in treating plant sites for more thorough protection of hearing.

In one of the new and more exotic fields of application, great forward strides will be made in glass fiber optics.

THE FUTURE OF FIBER GLASS†

The future is as bright and promising as it was 30 years ago. Major new uses are still being developed, both through improved economics of already established applications, and through new technologies.

The still-developing technology of manufacturing has allowed glass fibers to become a versatile, reliable, and low-cost raw material. It is made from widely available inputs and has a relatively low energy content. Its value as a reinforcement for elastomers and plastics has been well demonstrated; uses in these composites are just hitting their stride, and substantial growth is assured.

An example of a new technology is glass fiber reinforced cement which is on the brink of becoming one of the new major uses. The alkali-resistant grade of glass fiber that is required to reinforce portland cement has recently become readily available in commercial quantities, and the cement is abundantly produced throughout the

*L. R. White, Johns-Manville Sales Corporation
†Ralph H. Sonneborn, General Manager, Cem-Fil Corporation

world. This composite is a broadly useful engineering material that is remarkably compatible with the world of materials and energy shortages that are ahead of us.

BUILDING CODES*

The rapid increase in the use of plastics-based composite materials such as fiber-reinforced plastics attests the advantageous combination of strength, lightness, toughness, amenability to efficient forms, and fabricability of these materials.

Many of these applications involve load-bearing structures upon which life and safety depend. Too often they have been designed by haphazard or superficial methods revealing a lack of understanding of the peculiarities of these materials. Such approaches must be replaced by responsible professional design.

However, competent engineers and architects require reliable data and design procedures to guide them if they are to produce safe, economical designs. Methods applicable to traditional materials often are not suitable for structural plastics. Designers need codes and manuals to guide them, similar to the ones available for other structural materials such as concrete, metals, and wood. Such guides are largely nonexistent for plastics, but are even more needed for these unfamiliar materials than for the traditional materials already familiar to the designer.

This book should go at least part way toward that goal by providing much of the information about glass fiber reinforcement that the designer needs to allow him to design his structures intelligently. If it performs that task it will amply justify the effort that has gone into it.

PAST, PRESENT, AND FUTURE OF MOLDED INSULATING FIBER GLASS FOR AUTOMOTIVE APPLICATIONS†

Insulating fiber glass was first used in molded configurations for automotive applications in the early 1950's. Dash cushion pads and later molded hood-insulating pads appeared, the latter replacing die-cut fiber glass which had been used previously. These pads are acoustical

*Albert G. H. Dietz, Professor of Building Engineering, Massachusetts Institute of Technology
†Prepared by J. J. Clifford, Product Manager, Johns-Manville Sales Corp.

absorbers which prevent engine noise from entering the passenger compartment and they reduce the noise level perceived outside the vehicle. Fiber glass is an ideal material for this application because of its excellent acoustical absorption properties and its ability to withstand underhood temperatures. Because the phenolic resin-bonded material is inherently flame retardant, molded fiber glass does not have the underhood fire hazard potential of other fibrous automotive insulations. Molded hood insulation pads allow the automotive companies to cover the entire underhood area from fender to fender and dash panel to grill providing the maximum coverage for acoustical control. The molded configuration is designed to conform to the underhood contours and to clear engine accessories in a single easy-to-install pad. These pads are held in place with mechanical fasteners eliminating the costly and time-consuming adhesive bonding installation required with the original die cut fiber glass pads. Molded underhood pads are presently used in approximately 85% of all domestic automobiles.

Another molded insulating fiber glass product which appeared in the early 1960's is molded headliners. Insulating fiber glass headliners are supplied to the automotive companies prefinished with the facing of the individual company's choice. These headliners are both the interior trim for the vehicle and the acoustical roof insulation in one quick-to-install unit. Fiber glass headliners have gained acceptance in Detroit because they are far easier to install than the traditional cut-and-sew headliners which have been used in automobiles since the 1920's. The molded configuration is designed to conform to the contours of the roof, clearing roof bow supports when necessary and providing contours and recesses for dome lights, seat belt attachments, seat belt retractors, and sun visors. The unique characteristics of molded insulating fiber glass make it an ideal material for an automotive headliner substrate as it is virtually unaffected by atmospheric moisture, has excellent dimensional stability, is mildew and vermin resistant, and is inherently flame retardant. In addition, the material can be molded into virtually any configuration to provide stylists with design flexibility to meet headroom requirements or to provide highly styled headliners for speciality vehicles. Molded insulating fiber glass headliners are lighter in weight than competitive headliners made from hardboards. Lightweight is becoming one of the most important characteristics for automotive components as the automotive companies design much lighter and more fuel-efficient vehicles. Lightweight one-piece headliners will ultimately be used in virtually all domestic vehicles, many of which employ the molded insulating fiber glass substrate configuration.

The use of other molded insulating fiber glass components is growing rapidly as the potential for weight savings and the design versatility of the material are recognized and exploited by automotive engineers and designers. Miscellaneous molded components now include acoustical side cowl insulators, thermal floor insulators and foil faced van engine housing and cowl insulators. The van engine housing insulators provide acoustical and thermal insulation preventing discomfort inside the vehicle. The foil facing provides a metal septum which reduces ignition generated radio interference.

The automotive industry is currently looking at molded insulating fiber glass components to be used as dash panel insulators, cowl insulators, quarter panel insulators, and rear seat and floor insulators. Fiber glass is being considered for these applications for its acoustical and thermal insulating properties and often because it is lighter in weight than many of the insulations which are presently being used. In addition, the molded configurations simplify assembly by eliminating die cut multipiece assemblies.

The automotive market for molded insulating fiber glass has grown from its infancy in the early 60's to over 25 million pounds in 1977. In the 1980's, it is expected that this market will continue to grow and that molded components will find their way to a greater extent into recreational vehicles, light and heavy trucks, and farm equipment.

THE FUTURE FOR REINFORCED PLASTICS/COMPOSITES*

Prospects for long-term growth in markets for reinforced plastics/composites rest solidly on the physical properties of the materials and the economics of production. Proven materials can be expected to expand volume in accepted automotive, appliance, corrosion-resistant, construction, electrical, and other applications. New high-strength laminates with carbon fiber or high glass content are sure to take over many heavy-duty jobs from metals. Improved processing equipment and techniques will help make the materials competitive.

*Prepared by William H. Gottlieb, MORRISON-GOTTLIEB, INC., PUBLIC RELATIONS and ADVERTISING, NEW YORK, NEW YORK

Curiously, the same market factors that favor growth for RP composites also threaten to restrict it. Concern for the environment dictates lighter cars, putting a premium on the excellent strength/weight ratio of reinforced plastics; equipment for waste treatment, water and air purification makes increasing use of corrosion-resistant RP products. But many public officials and agencies tend to see all plastics as sources of pollution. Similarly, concern for the diminishing energy supply encourages use of RP's strength, light weight, heat insulation, and production efficiency. But the industry has to fight for its small share of oil and natural gas.

The reinforced plastics/composite industry has products uniquely suited to the needs of society, but it will achieve its potential only if it can marshal the evidence and present it convincingly to the general public and their elected officials.

FUTURE OF FIBER GLASS*

The result of 1976/1977 record cold has been greater emphasis on fuel conservation at the national level than we could have anticipated. When it was suggested that the fiber glass insulation manufacturers could face the need to double their capacity by 1980, we were not assuming that all inefficient, older and underinsulated houses would be insulated. That, however, is a possibility today. In fact, we had some concern last year about the fiber glass industry (only three producers have the necessary technology in the United States) adding capacity at a rate rapid enough to prevent shortages. Today those shortages appear imminent.

Two factors primarily are responsible, aside from the weather and the cost and shortages of fuel. Those are the development of "low energy" construction techniques for structures, including houses, and the favorable cost-payback ratio for added fiber glass insulation in older houses as the price of fuel rises. The Arkansas House was one of the first structures designed to reduce energy consumption sharply, while not increasing costs of construction. Since, variations on this theme have evolved in other parts of the country and the validity of such a concept appears proven. The second factor, that of a favorable cost-payback for adding insulation to existing houses is the result of fuel prices which have risen 400–600% during a period when the

*F. A. Winterhalter, Director of Research, Johns-Manville Sales Corp. Industrial Products and Glass Technology,

manufacturer's price for fiber glass insulation rose about 40%. Where two years ago, no economic justification could be advanced for insulating many older homes, today we can see a quick return of capital employed in such ventures.

The result of these dramatic changes is currently creating a soldout condition in the fiber glass insulation industry, which should persist through the next five years. The industry is not planning to add capacity for this surge in demand to new levels (based on higher standards and improvement of the existing housing stock) and then have demand decline drastically thereafter. Johns-Manville is on record that they are in a 5-year double capacity program. The difficulty of entry into fiber glass manufacture reinforces this view, since only three participants, Certainteed, Owens-Corning Fiberglas, and Johns-Manville, will be making the decisions to expand capacity.

FUTURE ADVANCES IN FIBERGLAS TECHNOLOGY*

Major technical advances in glass fiber technology for both continuous filament products for reinforcements and blown filament products for construction materials will be in two major areas: (1) energy reduction in the glass melting and fiber forming processes, and (2) improved product properties, particularly in high temperature properties and in product durability.

The increasing cost of energy and the increasing restrictions on energy availability are major incentives for the development of more energy efficient processes. These improvements will be achieved by the invention and development of entirely new concepts and by the examination and refinement of older known technology which becomes economically viable as the cost of energy increases.

Product property improvements have always been a major objective in the glass fiber industry. An accelerated rate of such advances is anticipated, however, as new and more energy efficient processes are developed which will provide opportunities for the introduction of new glasses with improved properties. Emphasis on higher temperature fibers for insulating materials will result from the need for such materials in many energy related processes and in energy conservation requirements.

Improved mechanical properties and chemical durability of reinforcement fibers will greatly increase the rate of replacement of con-

*Prepared by G. R. Machlan, Senior Research Fellow, Owens-Corning Fiberglas Corp.

ventional materials in structural applications by fiber reinforced composites. The transportation industry in general and the automobile in particular requires lower weight structural materials. Advanced glass fiber composites and hybrid composites containing glass fibers will meet many of these requirements and will find increasing applications in all areas where cost, performance, and weight are all critical to material selection.

THE FUTURE OF GLASS FIBERS*

Glass fiber technology, although already monumental in its breadth of achievement, appears still to be in an embryonic stage of development in relation to the potentials and knowledge unfolding in laboratories. The future of glass fibers in application, even in view of the over 50,000 uses estimated to exist, appears to be only at the base of the rising curve of development toward maturity.

New developments in thermal, acoustical, electrical, and structural applications including the diverse applications of resin composites are being announced continually. Many basic properties of glass fibers are still not fully understood and their applications are relatively unexplored.

Reinforcement of other matrices with which glass fibers might be combined is relatively untapped. Considering the possible reinforcement of metal alone, the capabilities of combining any one of the expanding spectrum of glass fiber products with a composition within the spectrum of any one of a number of metals are far from known.

Thus, the future of glass fibers from the standpoint of fundamental understanding is still open to many years of exploration. The unlimited availability of raw materials used in making glass fibers and the wide span of properties still unexplored provide a base for many more generations of research and development of glass fibers as a basic material.

FUTURE OF FIBER GLASS MAT AND STAPLE FIBER PRODUCTS†

When the present advanced state of the glass fiber mat and staple fiber industry is compared to its modest beginnings, we must believe that

*Prepared by Charles F. Schroeder, Senior Patent Counsel, Law Department, Owens-Corning Fiberglas Corp.
†Prepared by F. E. Schlachter, Division Manager, Production and Engineering, Europe, Johns-Manville Corp. Wertheim, Germany.

the future is with us now. Whereas lightweight glass fiber webs, or mats, were originally developed to overcome the severe difficulties experienced with rag felts used in bituminous built-up roof membranes, such as rotting, blistering and wrinkle cracking, etc., they are now also replacing asbestos and organic fiber combinations in roofing felts because of environmental problems.

Additionally, fiber glass webs, due to their many advantages, are finding use in a variety of reinforced plastics products such as light domes, telephone booths, mail boxes, oil tanks, chemical storage tanks, trailers, mobile homes, packaging media, and reinforcing agents for thermoplastics, asphalt, many types of foams, even in cement, and other products too numerous to mention. Product properties such as high tensile strengths, resistance to most chemicals, high dimensional stability, good weathering properties, and lightness of weight are imparted. In the case of organic plastics and foams, a substantial degree of fire retardancy is induced by the glass fiber.

These glass fiber webs are also bonded to the back of vinyl cushion material to form a superior flooring, used as an over-wrap on PVC pipe to impart strength and around steel pipe for corrosion protection. They are also laminated with wood veneers as plywood, and bonded with aluminium foil for facings. In each instance, the results over the original nonreinforced material are remarkably improved mechanical properties and superior dimensional stability.

Strandlike glass staple fiber, or sliver products, are used jointly with glass mats as battery separator membranes to provide the longest battery life of any presently known separator material. These textile-type materials, available in many deniers and strand configurations, also are converted into attractive woven patterns for fire retardant curtains, draperies, and wall decoration for select interiors.

Coarser industrial-type fabrics are also woven and used for roofing and for many types of laminating. In addition, single staple fiber strands are used as both edge and central reinforcing strands in glass fiber mat products for roofing, etc. Numerous nonwoven scrim type fabrics are also produced using the staple fiber strands.

Both mats and staple fiber strands are widely used in filtration applications. Staple fibers themselves have been processed into packings, ropes, insulation over-braided onto wire and tubing, and gasketing materials of a variety of shapes. Uses are for refrigerator and oven seals, glands and valve packings, automotive automatic chokes, etc., and many other applications in which the excellent high-temperature properties and resistance to deterioration make these glass fiber products highly desirable over competitive materials.

The processes for these mat and staple fiber materials have been licensed to many firms throughout the world, and it is anticipated that rapid growth and further innovation in this group of unique fiber glass products will continue.

References

1. Oleesky, S. S., and Mohr, J. G. *Handbook of Reinforced Plastics of SPI.* New York: Van Nostrand Reinhold Co., 1964, pp. 121–2.
2. Cure plate manufactured by Thermo-Electric Co., Cleveland, Ohio.
3. Close, P. D. *Sound Control and Thermal Insulation of Buildings.* New York: Van Nostrand Reinhold Co., 1966, pp. 27 and 140.
4. Technical Bulletin No. 1-AC-6655A, September, 1976, Owens-Corning Fiberglas Corp.
5. Pirn, R. Acoustical misconceptions in open planning. *Progressive Architecture*, August, 1972.
6. Occupational Safety and Health Administration Publication, Item 1910.95, p. 99, 1975.
7. Technical Communication, J. Verschoor, Johns-Manville Sales Corp. Research Division, Denver, Colo.; see also publication No. 1-IN-6964-A, Owens-Corning Fiberglas Corp., Toledo, Ohio.
8. Johns-Manville Sales Corp. Technical Bulletin No. MO-13A, 6/76, "HUD Mobile Home Construction and Safety Standards and the New Insulation Levels."
9. Zeston Division, Johns-Manville Sales Corp., Bulletin No. PVC-7-74.
10. Technical Bulletin No. IND-3203, 3/76, Johns-Manville Sales Corp.
11. Bulletin Code No. 30-31-49U- 1/76 3M, Certainteed Products Corp.
12. Technical Publication No. 3-UF-6072, 12/72, Owens-Corning Fiberglas Corp.
13. Bulletin No. IND-3104, 3/76, Johns-Manville Sales Corp.; see also Bulletin Code No. 30-31-52U, Certainteed Products Corp., and Publications No. 5-IN-7535, 8/76, and No. 4-IN-6056B, 8/74 (Computer Determination of Economic Thickness for Pipe Insulations), Owens-Corning Fiberglas Corp.
14. Publication No. 1-UF-6025A, 8/74, and corollary publications Nos. 3-UF-6000 to 6007, inclusive, 12/72, Owens-Corning Fiberglas Corp.
15. Technical Bulletin No. IND-3130, 2/76, "How To Use Heat-Flow Graphs," Johns-Manville Sales Corp.
16. Technical Bulletin No. IND-3211, 3/76, p. 14; No. FG 263A, 3/76; No. FG-202A, 3/76; No. FG-234A, 3/76; and No. FG-382A, 3/76, Johns-Manville Sales Corp.

17. Technical Bulletin No. IND-4001, 4/76, Zeston Division, Johns-Manville Sales Corp.

18. Technical Bulletin No. FG-297A, 3/76, Johns-Manville Sales Corp.

19. Technical Bulletin No. FG-328A, 3/76, "Fiber Grip Mechanical Closure System," Johns-Manville Sales Corp.

20. Bulletin No. IND-3235, 2/76, Air-Handling Systems, "Gump Solar Collector, Denver, Colo.," Johns-Manville Sales Corp.

21. Bulletin No. 1-MS-6431, 3/74, Owens-Corning Fiberglas Corp.

22. Bulletin No. IND-3171, 3/76, Johns-Manville Sales Corp.

23. Personal Communication, John Potter, Flight Insulations, Inc., Marietta, Ga.

24. Personal Communication, Robert White, Lockheed Missiles and Space Co., Inc. Sunnyvale, Calif.; see also brochure, "Thermal Protection and Control Systems," Lockheed Missiles and Space Co., Inc., October, 1972.

25. Publication on Fiberfrax® ceramic fiber, form A-2302-1/2, "Bulk Fiber," Carborundum Co.; see also Bulletin NV/111/1ED/752/274; ICI Mond Division.

26. Bulletin No. IND-3115, 12/73, Johns-Manville Sales Corp.; see also *Firewall® Installation Manual*, Carborundum Co., and Colson, F. A. *Kiln Building With Space-Age Materials.* New York: Van Nostrand Reinhold Co., 1975.

27. Bulletin No. A2299, Carborundum Co.

28. "Ceramic Fiber vs Insulating Brick—A study," Clark, R., Johns-Manville Sales Corp., in *Heating/Combustion Equipment News*, Dec./Jan. 1974, p. 1.

29. WRP Bulletin, Refractory Products Co.; see also Technical Publication No. A-2309-1, Carborundum Co.

30. Tech Brief No. 67-10607, NASA.

31. Report No. N66-13572, NASA.

32. Hoppit, H. B. Testing and application of filters for air-conditioning. *Filtration and Separation*, Nov./Dec. 1974. See also Holzauer, R. Selecting air filters. *Plant Engineering*, March 21, 1974, p. 6; and Farrow, R. M. The filtration characteristics of glass fiber media. *Filtration and Separation*, Nov./Dec. 1966.

33. McGowan, J. P. Air filtration in the building services industry. *Filtration and Separation*, Nov./Dec. 1975, p. 684.

34. Personal Communication, D. C. Perry, Johns-Manville Sales Corp.

35. ASHRAE Standard No. 52-68, "Method of Testing Air-Cleaning Services Used In General Ventilation for Removing Particulate Matter," August 19, 1968; see also ASHRAE 1969 *Guide and Data Book*, Chapter 10, p. 107.

36. Personal Communication, John J. Preast, Filter Fabricators, Inc.

37. Liquid filtration test dust supplied by AC Division, General Motors Corp.

38. *Plant Operating Management*, July 1971, p. 43.

39. Technical Bulletin No. MD 16 A, "Membra-Fil Membrane Filtration Technology," Johns-Manville Sales Corp.

40. "Glass, Ceramic, and Quartz Fiber for the Paper Industry," N. B. Scheffel, Johns-Manville Sales Corp., *TAPPI Journal* Vol. 58, Nov. 5, May 1975, p. 56.

41. "Reports on Ten Years of Roofing Research, "Fourth Edition, 1963, Midwest Roofing Contractors' Association, Kansas City, Mo.; see also "Here's How to Make a Roof Drain," Ibid. May 1972.

42. "Dimensional Stability and Shape Retention in Floor Coverings," Dr. Ing. W. Körtje, VDI, Textil Betrieb, Heft 9, September 1973.

43. Oleesky, S. S., and Mohr, J. G. *Handbook of Reinforced Plastics of SPI.* New York: Van Nostrand Reinhold Co., 1964, p. 121.

44. Personal Communication, George C. Stover, Gould Battery Co. (Inventor of mat-sliver mat combination for lead-acid battery retainer).

45. Strickland Processes, U.S. Patent Nos. 3,573,014; 3,905,790; 3,981,704; and 3,986,853.

46. Jaray, F. F. "A New Method of Spinning Glass Fibers," 28th Annual SPI RP/C Institute Proceedings, 1973, Section 3-A.

47. Technical Communication, Galileo Electro-Optics Corp., Sturbridge, Mass.

48. Bortz, S. A., and Li, P. C., "Continuous-Filament Refractory Fibers," 22nd Annual SPI RP/C Institute Proceedings, 1967, Section 7-C.

49. Oleesky, S. S., and Mohr, J. G. *Handbook of Reinforced Plastics of SPI.* New York: Van Nostrand Reinhold Co., 1964, p. 121-2.

50. Mohr, J. G. *et al. SPI Handbook of Technology and Engineering of Reinforced Plastics/Composites.* New York: Van Nostrand Reinhold Co., 1973, pp. 30-51.

51. Personal Communication, C. Pedersen, Fiber Glass Industries, Amsterdam, N.Y., and J. N. Grove, Professionel at Rossmoor, Jamesburg, N.J.

52. U.S. Design Patents Nos. 229,672 to 229,686, inclusive, by Mark L. Hildebrand, Altadena, Calif., assignor to Polyarch Homes, Division of Rudkin-Wiley Corp., Seymour, Conn.

53. Mallinson, J. H. "Chemical Plant Design with Reinforced Plastics," McGraw-Hill, Inc., 1969.

54. Hamner, N. E. (ed). *Non-Metals Corrosion Data Survey*, 5th ed, Houston, Texas: National Association of Corrosion Engineers, 1975.

55. Fuller, A. B. U. S. Patent No. 3,524,422.

56. Product Data Bulletin No. IND-3108, 3/76, Johns-Manville Sales Corp.

57. Nelson, N. B., Selecting filter fabrics for bag houses. *Plant Engineering*, October 28, 1976, File No. 7550; see also brochure "Fabrics for Filtration and Dust Collection," J. P. Stevens Co.

58. Bulletin No. 1-IN-6586A, Oct. 1974, Owens-Corning Fiberglas Corp.

59. Technical Communication, A. J. Rivard, 3M Company, 11/23/76.

60. Bulletin No. 5-GT-5442-A, 3/72, "Textile Fibers for Industry." See also nomenclature data, p. 13 (or equivalent publication).

61. *Insulation Circuits*, November, 1975, p. 36; see also Annual Report for 1976, M.B. Associates, San Ramon, Calif.

62. "Structural Properties and Moisture Resistance of Surface-Bonded Concrete Masonry Walls," Research Report by National Concrete and Masonry Association Research Laboratory, for Owens-Corning Fiberglas Corp., March 1971, et seq.; May 1971; see also ACI Journal Abstracts, August 1974, SP-44-21, p. 418.

63. CEM-FIL Technical Bulletins, CEM-FIL Corp., Nashville, Tenn., personal communication, Ralph Sonneborn.
64. U.S. Patent No. 2,184,326, Thomas, J. H., assigned to Owens-Corning Fiberglas Corp.
65. Personal communication, M. Crouther, Owens-Corning Fiberglas Corp., April 7, 1977; see also Wolf, R. F. Radial vs belted bias ply tires. *Rubber Age*, April, 1969.
66. *Plastics Technol Mag*, November, 1976, p. 37.
67. U. S. Patent Nos. 2,589,792; 2,923,113; and 3,624,385.
68. U.S. Patent No. 3,758,285.
69. "Fiber Optic Principles, Properties, and Design Considerations," W. P. Siegmund, American Optical Co.
70. Technical communication, Galileo Electro-Optics Corp., Sturbridge, Mass., their Technical Memorandum No. 100.
71. Technical communication, Bell Laboratories, Norcross, Ga, 3/10/77.
72. Paper Summary No. Th B6-1, "Optical Fiber Transmission II," Conference Papers of Optical Society of America Meeting, Feb. 22–4,1977, Williamsburg, Va.
73. Ibid, Section Th B3-1, "Optical Systems for (Japanese) Electric Power Companies."

Index

329